# PHILOSOPHY AND CLIMATE SCIENCE

There continues to be a vigorous public debate in our society about the status of climate science. Much of the skepticism voiced in this debate suffers from a lack of understanding of how the science works – in particular the complex interdisciplinary scientific modeling activities such as those which are at the heart of climate science. In this book Eric Winsberg shows clearly and accessibly how philosophy of science can contribute to our understanding of climate science, and how it can also shape climate policy debates and provide a starting point for research. Covering a wide range of topics including the nature of scientific data, modeling, and simulation, his book provides a detailed guide for those willing to look beyond ideological proclamations, and enriches our understanding of how climate science relates to important concepts such as chaos, unpredictability, and the extent of what we know.

ERIC WINSBERG is Professor of Philosophy at the University of South Florida. He is the author of *Science in the Age of Computer Simulation* (2010) and has published in a number of philosophy journals including *Philosophy of Science, The Journal of Philosophy, The British Journal for the Philosophy of Science, Studies in History and Philosophy of Modern Physics*, and *Synthese*.

# PHILOSOPHY AND CLIMATE SCIENCE

ERIC WINSBERG

*University of South Florida*

CAMBRIDGE
UNIVERSITY PRESS

# CAMBRIDGE
## UNIVERSITY PRESS

University Printing House, Cambridge CB2 8BS, United Kingdom

One Liberty Plaza, 20th Floor, New York, NY 10006, USA

477 Williamstown Road, Port Melbourne, VIC 3207, Australia

314–321, 3rd Floor, Plot 3, Splendor Forum, Jasola District Centre, New Delhi – 110025, India

79 Anson Road, #06-04/06, Singapore 079906

Cambridge University Press is part of the University of Cambridge.

It furthers the University's mission by disseminating knowledge in the pursuit of education, learning, and research at the highest international levels of excellence.

www.cambridge.org
Information on this title: www.cambridge.org/9781316646922
DOI: 10.1017/9781108164290

First published 2018

Printed in the United Kingdom by TJ International Ltd. Padstow Cornwall

A catalogue record for this publication is available from the British Library.

Library of Congress Cataloging-in-Publication Data
Names: Winsberg, Eric B., author.
Title: Philosophy and climate science / Eric Winsberg, University of South Florida.
Description: New York: Cambridge University Press, 2018. |
Includes bibliographical references and index.
Identifiers: LCCN 2017060367 | ISBN 9781107195691 (hardback ) |
ISBN 9781316646922 (paperback )
Subjects: LCSH: Climatic changes – Social aspects. | Science – Social aspects.
Classification: LCC QC903.W587 2018 | DDC 363.738/7401–dc23
LC record available at https://lccn.loc.gov/2017060367

ISBN 978-1-107-19569-1 Hardback
ISBN 978-1-316-64692-2 Paperback

We are now faced with the fact that tomorrow is today. We are confronted with the fierce urgency of now. In this unfolding conundrum of life and history there is such a thing as being too late. Procrastination is still the thief of time. Life often leaves us standing bare, naked and dejected with a lost opportunity. The "tide in the affairs of men" does not remain at the flood; it ebbs. We may cry out desperately for time to pause in her passage, but time is deaf to every plea and rushes on. Over the bleached bones and jumbled residue of numerous civilizations are written the pathetic words: "Too late."

Martin Luther King, Jr. – Speech delivered April 4, 1967

# Contents

# Figures

# Boxes

# Tables

# *Preface*

The philosophy of climate science is a sub-field of the philosophy of science that is growing very fast. Just a few years ago, there were only a few people who had published on this topic, but now there are dozens. There are also a growing number of aspiring philosophers in graduate school who are turning their attention to climate science as an area of science that is both extremely socially relevant and ripe for philosophical investigation. A few well-known climate scientists, including Reto Knutti, Gavin Schmidt, Leonard Smith and Jonathan Rougier, have even written papers that are explicitly intended as contributions to the philosophical literature.

The goal of this book is both to provide an introduction to this growing literature to those interested in what philosophy of science can contribute to our understanding of climate science and its role in shaping climate policy debates, as well as to advance the debate on many of its topics. The first four chapters of the book are more or less introductory, and should be accessible to anyone regardless of their background in either climate science or philosophy. The remainder of the book builds on this background. After reading the first four chapters, each of the following sections is more or less self-contained: Chapter 5 (plus the appendix), Chapters 6–9, Chapters 10–12, and Chapter 13.

I'm in debt to a number of people and institutions for helping me make this book possible. I'll try to mention a few of them here. Much of the book was written while on a fellowship at the MECS at Leuphana University, Lüneburg, funded by the Deutsche Forschungsgemeinschaft (DFG) in the spring of 2016 and while on sabbatical at the University of South Florida in the fall of 2016. I want to thank Roger Ariew, my department chair, for always making it easy to get time to work on the book – especially during those two semesters. Thanks to Martin Warnke and Claus Pias not only for inviting me to the MECS but for making the whole thing happen in the first place. And thanks to Jantje Sieling for all her tireless work in helping me, Jessica, and Chora have such an easy time visiting Lüneburg.

And thanks to Mathias Frisch and Torsten Wilholt for giving us a nearby second home in Germany with lots of good philosophical conversation.

I want to thank Hilary Gaskin at Cambridge for her early encouragement and for her skillful shepherding of the book from its earliest stages up into production and Colette Forder for her patient and careful copyediting. Lots of people have read parts or all of this manuscript and have given me valuable feedback. I especially need to thank Mathias Frisch and Elisabeth Lloyd for this, but also Wendy Parker, Doug Jesseph, Dasha Pruss, and Jonah Schupbach. But nobody has looked at the manuscript quite as carefully as Lukas Nabergall and Alejandro Navas – the two best students anyone could ever hope to have. Discussions with them were especially helpful, as any reader will be able to see, in understanding exactly what the relationship is between chaos, non-linearity, and structural stability. They also caught literally hundreds of errors, ranging from minor typos to mathematical mistakes. Thanks also to both of them, as well as to Wendy Parker, for letting me borrow freely from joint work in writing the appendix, and Chapter 9, respectively.

Thanks most of all to Jessica Williams, who pushed me to write this book in the first place, who read every draft of not only every chapter, but of the proposal, the emails to important people, and everything in between. And thanks to her, all the more, for her love and support all through the whole process of making the book come into being. Finally, though she never cared a whit about books, or philosophy, or climate science, or really anything other than chasing balls, sticks, rabbits, and squirrels, I need to thank Chora for always getting me out of my head when I needed a break from the project.

# 1

# *Introduction*

"Of course not every single adherent of the scientific world-conception will be a fighter. Some glad of solitude, will lead a with-drawn existence on the icy slopes of logic." From the Vienna Circle's manifesto, *Wissenschaftliche Weltauffassung*
(Hahn, Neurath and Carnap 1929)[1]

2016 was the warmest year on record.[2] It broke the record of 2015, which broke the record of 2014. The nine consecutive months from December 2015 to August 2016 were all record-breakingly warm. This was the fifth time in the twenty-first century that a new record had been set. All 16 of the years that have passed in the twenty-first century are among the 17 warmest on record (with 1998 rounding out the lot.) All five of the five warmest have been since 2010.

Regionally, the patterns have been a bit more complicated but reflect the underlying trend. Ocean surface temperatures had their warmest year; all six continents experienced one of their top five warmest years on record, and Arctic sea ice experienced its smallest seasonal maximum ever and its second smallest seasonal minimum.[3] The 13 smallest seasonal maximums have all been in the last 13 years. The melting of Arctic ice is an especially significant change in the global climate because of its feedback effect: as the temperature rises, ice melts, and melting ice reduces the amount of sunlight reflected back into space, which makes the temperature rise

---

[1] Translated in Neurath and Cohen (1973), p. 317, quoted in Reisch (2007), p. 58

[2] The average global surface temperature for 2016 was 0.94°C above the twentieth-century average, 0.83° celsius above the long-term average (14°C) of the World Meteorological Organization (WMO) 1961–1990 reference period, and about 0.07°C warmer than the previous record set in 2015. This is approximately 1.1°C higher than the pre-industrial period.

[3] Peak ice in the Arctic in 2016 was reached on March 24 at 5.607 million square miles (14.522 million square kilometers) – the smallest peak ever. 2012 still barely holds the record for the lowest summer minimum area of the Arctic Ocean covered by ice with a low of 1.58 million square miles (4.1 million square kilometers).

even more. The melting of the Arctic permafrost, moreover, could release billions of tons more carbon and methane into the atmosphere – also accelerating warming.

Precipitation patterns continued to get more extreme, with some regions experiencing record drought (especially southern Africa and Australia) and some experiencing record flooding (especially China and Argentina). Some regions experienced both.[4] Understanding and predicting the impact of warming temperatures on regional precipitation remains a serious challenge.

Climate change is real, and it is happening in front of our eyes. And while Americans are almost evenly split with regard to whether or not they believe that human activities are the cause of these changes, the scientific community is not. The relevant experts on the climate system are virtually unanimous in their acceptance of anthropogenesis: the proposition that human activities (primarily in the form of the combustion of fossil fuels, but also the extraction of those fuels, deforestation, livestock farming, and the manufacture of concrete) are responsible for at least the bulk of those changes. Not only do climate experts unanimously hold these views. So do virtually all the members of neighboring scientific disciplines and their scientific societies.

Still, well-meaning people sometimes conflate that unanimity with the idea that anthropogenesis is an obvious truth – that it can be established with ease or simplicity. I was at a public lecture once where the speaker (a journalist) said that the truth of anthropogenesis was like 1+1=2. I do not think this kind of rhetoric is helpful. One obvious danger of overstating the simplicity of the reasoning is that it encourages poorly informed laypersons to think they can evaluate the reasoning themselves, and potentially find simple flaws in it. It's true that the greenhouse effect, which explains why the earth doesn't look like the ice planet of Hoth in *The Empire Strikes Back*, is a simple mechanism involving heat-trapping gasses, and one whose existence is easy to establish. And it's not hard to show that humans have been producing ever-growing quantities of those gases for over 200 years. So, there is some relatively simple reasoning, based on a simple model,[5] that makes the hypothesis of anthropogenesis plausible and perhaps even more likely true than not. But the community of experts believes unanimously in anthropogenesis not merely because of this

---

[4] The United Kingdom, for example, had record-setting rainfall in December/January 2015–2016 and then a bottom quartile autumn accumulation.

[5] See section 3.2

simple reasoning, or because of anything that should be likened to 1+1=2. They believe it because of decades of painstaking work in collecting and studying data, pursuing multiple independent lines of evidence, building and studying complex models run on clusters of powerful computers, and recruiting into their ranks the expertise of literally dozens of different scientific disciplines: Climatology, Meteorology, Atmospheric physics, Atmospheric chemistry, Solar physics, Historical climatology, Geophysics, Engineering, Geochemistry, Geology, Soil science, Oceanography, Glaciology, Paleoclimatology, Ecology, Synthetic biology, Biochemistry, Biogeography, Human geography, History, Economics, Ecological genetics, Applied mathematics, Mathematical modeling, Computer science, Statistics, and Time series analysis, just to name a few.[6]

In short, the scientific study of the climate and its response to human activities isn't just vitally important to the future of the planet. It's also rich, interesting, complex, and deeply interconnected with almost every area of study that occupies the minds of twenty-first-century scientists. On top of all that, it is literally awash with all the conceptual, methodological, and epistemological issues that perennially preoccupy philosophers of science: the nature of scientific data and its relation to theory; the role of models and the role of computer simulations in the practical application of theory; the nature of probabilities in science and in decision making; how to think about the latter when the probabilities available seem ineliminably imprecise; the methodology of statistical inference; the role of values in science; confirmation theory; the role of robust lines of evidence in confirming hypotheses; social epistemology (the value of consensus in science; group knowers and authors; the value of dissent) and too many others to list.

It's just the kind of scientific practice that you would expect philosophers of science to take an exceptionally keen interest in. But until relatively recently, you would have been pretty disappointed. The reasons for this are complicated. One reason is that philosophers of science tend to cluster around a small group of scientific topics, in which they collectively build expertise, and about which there is collective agreement that they are "philosophically interesting." A Martian, visiting earth, who tried to learn about the range of scientific activities in which humans engage by visiting a meeting of the Philosophy of Science Association, would find us to be very parochial in our interests.

---

[6] (Brook 2008)

Another reason might have to do with philosophy of science's withdrawal to "the icy slopes of logic" during the post-war McCarthyite period of American academic history, detailed by George Reisch, Don Howard, and others.[7] Reich and Howard remind us of a time when American philosophers of science followed the leadership of the members of the Vienna Circle (who had come to the United States to flee the Nazis), not only with respect to their epistemological and (anti)metaphysical commitments, but also with respect to one of their deepest motivations: that philosophy of science should be engaged with "the life of the present," and pursue the aim of turning the scientific enlightenment toward the project of bettering the social conditions of mankind. But the pre-war association of those same philosophers with workers' parties and democratic socialism put the careers of their followers in peril. In reaction, the general character of philosophy of science in the English-speaking world became politically neutralized: distanced from issues of social concern, and focused on areas of science of little social consequence.[8]

Whether in part because of the warming of the climate or not, and certainly in no small part due to the growing influence of feminist philosophy of science, the icy slopes of logic have been melting of late, and the number of philosophers of science interested in socially relevant philosophy of science has grown in the last decade or two. Socially relevant philosophy of science can mean a variety of different things,[9] but it certainly refers to philosophical work that engages with science that has significant social impact. It is therefore no surprise at all that there is a growing interest in climate science among philosophers of science of late – both as a research topic in its own right, and as a useful case study that is easily adaptable to philosophy of science pedagogy. The topic also complements much of the recent work on climate ethics. Climate ethics is primarily concerned with ethical issues that surround climate change and how issues of justice bear on the duties and responsibilities producers of greenhouse gases have toward those they will affect. This work is best done in the context of a reasonably good understanding of the science of climate change, and thus climate ethicists can certainly benefit from a philosophically informed presentation of the foundations of climate science.

---

[7] (Reisch 2007, Howard 2003)

[8] The earliest exceptions to this were probably when feminist philosophers of science started (in the 1980s or so) to take an interest in areas of biology and social science that had implications for gender-related social concerns. "Socially relevant philosophy of science" and "feminist philosophy of science" were for a relatively long period virtually synonymous.

[9] See Fehr and Plaisance (2010)

This book was written for the benefit of everyone who wants to come down from the icy slopes, as well as for climate scientists curious about what philosophers think about their work. I hope for it to serve both as an introduction to the major themes of the philosophy of climate science, and as an effort to add to that enterprise – to advance our philosophical understanding of the field. It is written to be as useful as possible, in the first instance, for students and scholars in philosophy of science who are interested in exploring climate science as a topic of philosophical study. But it is also intended to be accessible to a wider general audience, and to be useful as a resource for people studying general philosophy of science who prefer to see that material presented with real, living examples of scientific practice. I certainly hope some climate scientists will be curious to see what philosophers think of their discipline.

The book is not intended to be a polemic in defense of climate science or in defense of anthropogenesis.[10] Almost everywhere, I will be assuming that, with regard to questions about which the community of climate scientists share broad agreement, the answers that climate science delivers are the best answers we can find. I will be primarily interested in interpreting those answers (when it isn't obvious how to do so) and uncovering the logic, methodology, and conceptual foundations of the reasoning used to produce those answers.

The first part of the book is primarily about the methodology of climate science: Chapter 2 is about climate data and the relations between those data and climate hypotheses. Chapter 3 is about climate models in general, with an emphasis on static, equilibrium models of global radiation balance. Chapter 4 is on climate simulations. Chapter 5 is on chaos and its implications for climate science, particularly with regard to the difference between making predictions and making projections, and the nature of a "forcing experiment," which is one of the main ways in which simulations are used in climate science.

The second part of the book is mostly about uncertainty, and about how to interpret climate hypotheses for which we have only probabilistic support: Chapter 6 is on the interpretation of probability in climate science. Chapter 7 is on the related notion of "confidence" in climate projections, and on the nature and origins of climate uncertainties. Chapter 8 is on statistical inference and on decision making under

---

[10] Readers interested in more polemical defenses of climate science, and in particular in works that are primarily directed at refuting the arguments of climate skeptics and climate deniers, can turn to many good such resources that are already out there.

uncertainty and decision making under risk. It includes a discussion of so-called integrated assessment models, which try to make decision making itself model-based and scientific. Chapter 9 is on the interplay between uncertainty quantification in climate science and the role of social values in climate science.

The last part of the book is mostly on epistemological issues: Chapter 10 is on evaluating model skill, including discussions of "verification and validation" of climate models and of the epistemological impact of the fact that climate models are "tuned." Chapters 11 and 12 are both on the role of "robustness analysis" in climate science: that is, on the epistemological importance of the fact that some climate hypotheses are supported by a variety of lines of evidence, and of the fact that some hypotheses are jointly predicted by a whole ensemble of different models. Chapter 13 is on the application of various themes from "social epistemology" to the epistemology of climate science. Chapter 14 offers some concluding remarks.

# 2

# *Data*

## 2.1 Introduction

When we think of climate science, many of us think of the massive earth system simulations that are run on supercomputers. Even though climate *modeling* gets most of the attention, the important claims of climate science are also supported by data from many different and independent sources. We will start by discussing the observations and data that support the claim that the climate has changed significantly in the recent past. We can divide these into at least seven different categories:

1) Records of scientific instruments of various kinds that measure the local surface temperature of the earth and other closely related variables
2) Glacier records
3) Sea ice records
4) Ice sheet records
5) Ocean heat content records
6) Sea level records
7) Measured proxies for climate variables from times prior to the existence of measuring instruments (tree rings, ice cores, bore holes, pollen, corals, ocean sediments, etc.)

## 2.2 The Nature of Scientific Data

In discussing the data that support climate hypotheses of all kinds, it is important to keep in mind a recurring theme in empiricist-minded philosophy of science of the last century, which can be traced back at least to the work of Pierre Duhem: the theory-ladenness of data (see Box 2.1). Early twentieth-century empiricist philosophers of science, whether they were falsificationists, inductivists or hypothetico-deductivists, hoped that hypotheses could be tested against objectively secure, raw observations.

It has long been recognized, however, that hypotheses are only inferred from, and tested against, data whose reliability has been secured by a variety of complex and many-layered means. Raw data are regimented and restructured into an idealized pattern or set of patterns, what Patrick Suppes called a "*model of data*"(Suppes 1969), before they are used to generate and evaluate hypotheses.

Techniques for transforming raw data into a model of the data include, but are not limited to: statistical inference, error and noise reduction, the elimination of confounds, and understanding and modeling the behavior of measuring instruments. Motivating and justifying these techniques, in turn, often requires trust in other theories and hypotheses, including those that might themselves be supported by equally complex chains of reasoning. Again following Suppes, we can call these appeals to auxiliary theories and hypotheses that support the inference from raw data to models of data *models of the experiment* and/or *models of instruments*. Anytime we construct an experiment to collect data, we need models to tell us how to interpret what we find. And even the most mundane instrument needs to be well understood – we need a good model of how it behaves and responds to its environment – before we can make good use of the numbers it delivers. Do not suppose that I am exaggerating when I say "even the most mundane instrument." One controversy about sea temperature records was eventually resolved when researchers realized that the *kind of bucket* that was used to collect the water for temperature measurement had undergone a change in the 1940s that made a significant difference – the same water of the same temperature measured out of each kind of bucket would yield a significantly different measurement. Not surprisingly, when most historical climate data were collected no one envisioned that they would be used to try to reconstruct global, century-long records, and little attention was paid to collecting data that would be intercomparable with other data. Data collection practices varied, and this included variation in the kind of buckets used to sample water for temperature readings. Simple wooden buckets, of the kind you might picture if you were picturing a bucket hanging over a well, were used in the nineteenth century. Later on, special canvas buckets were used, and finally they used insulated buckets that don't look anything like what you probably picture when you think of a bucket. The effect that each of these buckets has on temperature readings can be significant – with the canvas buckets producing readings that are up to 1°C cooler than other buckets (Schmidt 2008).

Gavin Schmidt makes the point nicely vis-à-vis temperature records in a recent (2015) blog post on Realclimate.org:

The first thing to remember is that an estimate of how much warmer one year is than another in the global mean is just that, an estimate. We do not have direct measurements of the global mean anomaly, rather we have a large database of raw measurements at individual locations over a long period of time, but with an uneven spatial distribution, many missing data points, and a large number of non-climatic biases varying in time and space. To convert that into a useful time-varying global mean needs a statistical model, good understanding of the data problems and enough redundancy to characterize the uncertainties (Schmidt 2015).

In other words, the pattern of global mean surface temperature needs to be carefully reconstructed and inferred from the raw data of thermometer recordings and other records. We should also keep in mind that the relevant *thing* against which climate hypotheses are tested are not raw thermometer readings, satellite data, etc., but the structural patterns that can be inferred from the above – *the model of the data* – by the best available methods of inference and by the most responsible experimental and observational reasoning possible. Lay people (and I include philosophers of science in that category) are not always well-positioned to decide what those methods are. Sometimes, moreover, it is the very fact that there is a conflict between our working hypothesis and what we first take to be the best model of the data that spurs us to change our minds about what data model the raw data best support. As long as it is the most responsible pattern of reasoning that leads us down this path, data can even be laden with the very theories they support, and this needn't be viciously circular. We will see a nice example of this below. In cases such as these, whether or not those modifications count as "ad hoc," and as such provide legitimate grounds for criticism of the relevant inferences and conclusions, will depend entirely on the details. It requires, in the words of Pierre Duhem, not only "*bon sens*" or "good sense" but also meticulous and responsible reasoning to determine when this kind of apparent circularity is vicious and when it is virtuous.

## 2.3 Evidence of Warming

### *1) Records of Instruments*

Since at least the 1880s, meteorologists have been building meteorological stations to collect local weather data from points around the globe. The main phenomenon that all of these data support, directly and indirectly, is the existence of the warming pattern in the *global mean surface temperature*

---

**Box 2.1 The Duhem Problem**

The Duhem problem (or sometimes the Quine/Duhem problem, or Quine/ Duhem thesis) is the claim that it is impossible to test a scientific hypothesis in isolation, because interesting scientific hypotheses do not usually make direct contact with the world, and are often capable of making specific empirical predictions only with the help of a bundle of "auxiliary hypotheses." (For example, think of the hypotheses you need to postulate about the insulation around the post-war buckets, and how they affect temperature readings for sea-water.) Duhem himself argued for this thesis only with respect to physics, and he was primarily concerned with falsificationism (i.e. in showing that data could never falsify a physical theory by itself, since the failure of a theory to make a correct prediction could be blamed either on the theory or on one of the auxiliary hypotheses). Duhem argued that when the prediction of a theory conflicted with measured data, only "good sense," rather than logic, could determine where to place the blame and hence whether to reject the theory. Quine and Kuhn, among others, argued that the problem is more general. A slightly stronger thesis (which goes back primarily to Patrick Suppes), advocated in more recent philosophy of science, says that scientific hypotheses do not really confront raw data at all, but rather what Suppes called "models of data," or, similarly, what Woodward and Bogen called "phenomena"(Bogen and Woodward 1988).

---

that is depicted in Figure 2.1. Climate scientists usually measure warming in terms of what they call the *temperature anomaly*, a departure from a standard reference temperature. The reference temperature is usually taken to be the average temperature during the period from 1950 to 1980. A temperature that matches this average is stipulated to be zero degrees of temperature anomaly.

One argument that is often raised by skeptics of climate change is that this graph reflects (at least in part) what is known as the *urban heat island effect*. An "urban heat island" is a city or metropolitan area that is warmer than its surroundings because of human activities other than the emission of heat-trapping gases. The most obvious source of urban heat is the use of construction materials that store solar radiation; but there are others. Critics sometimes express the worry that, because many meteorological stations are close to cities, the effects of urban sprawl might have been misinterpreted as a rise in global temperature. But careful reasoning about which model of the experiment and which models of the instruments are best supported by the evidence can resolve this issue rather convincingly. The Third Assessment Report of the IPCC (see Box 2.2) provides part of the argument:

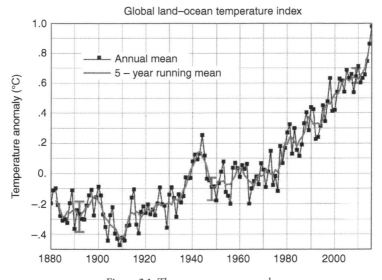

Figure 2.1 The temperature anomaly.
*Source*: adapted from http://data.giss.nasa.gov/gistemp/graphs_v3/ The vertical
bars show uncertainty estimates for selected years

Over the Northern Hemisphere land areas where urban heat islands are
most apparent, both the trends of lower-tropospheric temperature and
surface air temperature show no significant differences. In fact, the lower-
tropospheric temperatures warm at a slightly greater rate over North
America (about 0.28°C/decade using satellite data) than do the surface
temperatures (0.27°C/decade), although again the difference is not statistically significant. (IPCC, TAR, WG1, Box 2.1)[1]

In other words, if the heat island effect were responsible for the apparent
warming trend, then we would expect it to be larger where the cities are
most concentrated. But it isn't. In fact, the small difference that there is
(though it's too small to suggest anything with statistical significance)
points in the opposite direction.

## 2.4 Satellite Data: A Cautionary Tale

While data from meteorological stations go back to the late nineteenth
century, we also have data from satellites going back to the 1960s. Since
1978, satellites carrying microwave sounding units (MSUs) have been
measuring how much microwave radiation is being produced by oxygen

[1] For a thorough discussion, see Peterson (2003).

---

**Box 2.2 The IPCC**

The Intergovernmental Panel on Climate Change (IPCC) is an intergovernmental body set up to produce reports on state-of-the-art climate knowledge to support the United Nations Framework Convention on Climate Change. The IPCC does not carry out its own research. Rather, it provides assessments of the best knowledge contained in the published literature. To date, the IPCC has produced five "Assessment Reports" (plus one supplementary report). They are best known by their famous acronyms, FAR (first assessment report, etc.), SAR, TAR, AR4, and AR5. Each of the recent assessment reports is associated with an output of the "Coupled Model Intercomparison Project" – a project for summarizing the state-of-the-art output from climate models. Somewhat confusingly, until the latest round the numbering of CMIPs and the assessment reports were out of phase, with, e.g. CMIP3 supporting AR4. In the latest round, we jumped from CMIP3 to CMIP5 so that CMIP5 would be in phase with AR5.

---

molecules in the atmosphere, which bears an exploitable relationship to the temperature of broad layers of the atmosphere. Satellite data primarily concern the temperature distribution of the troposphere and lower stratosphere, but they can also record surface temperatures (of both land and sea) from infrared data in cloud-free conditions. Unfortunately, satellite data have generated even more controversy than ground-level records and illustrate at least as well the point about the theory-ladenness of data. Elisabeth Lloyd has made the point, moreover, that some of the controversies surrounding satellite data (and from closely connected data from radiosondes, or weather balloons) also illustrate the fact that the most reliable raw data are not always the ones that seem, from a naive point of view, to be mostly *directly tied* to the phenomenon of interest (Lloyd 2012). She uses the satellite example to advocate for what she calls "complex empiricism" which she contrasts with "direct empiricism," according to which the best hypothesis is always the one that most directly and easily reproduces raw data. From our point of view, we can use her study as a cautionary tale regarding the complexities involved in evaluating a hypothesis in the light of data. The moral will be that consumers of scientific information should beware of "experts" who peddle simple-minded arguments about what the data show.

The story (which is carefully documented in Lloyd's paper)[2] concerns the actual history of the distribution of temperatures from 1978 until

---

[2] What follows is a brief sketch of the episode and of Lloyd's contrasting notions of complex empiricism and direct empiricism. Readers interested in a more thorough account should be sure to consult the original (Lloyd 2012).

around 2000, when the controversy erupted. The controversy revolved around what the measurements revealed about the temperature record of the tropical troposphere. The troposphere is the lowest layer of the earth's atmosphere; in the tropics, it is about 20km deep. Despite the fact that, in principle, the troposphere includes the atmosphere all the way down to the surface of the earth, the earth's surface temperature (in any particular geographic region) is a distinct measurable quantity from the temperature of the troposphere. The controversy arose as a result of an apparent clash between the measured surface temperature in the tropics and that of the tropical troposphere. The latter was primarily estimated using satellite data. Global climate models had predicted that the heat-trapping gasses in the atmosphere would cause both the tropical surface temperatures, as well as the temperatures of the tropical troposphere, to rise – with the latter rising slightly faster than the former. But according to one very prominent group's model of the satellite data, the so-called UAH (for University of Alabama at Huntsville), the troposphere was not warming at all. It showed perhaps even that the tropical troposphere was cooling. This led UAH-affiliated climate scientists John Christy and Roy Spencer to conclude that *climate models were unreliable.*

If you have read Box 2.1, then you know that when the predictions of a model fail to match observed data, there are two possibilities. The model could be wrong, or the data could be wrong. In Duhem's terms, the auxiliary hypotheses used to bring the data into contact with the hypotheses could be false. In Suppes' terms, the model of the data (what climate scientists call the "data set") could be flawed. For many climate scientists at the time, the second possibility was no idle worry of philosophers. The temperature records of the troposphere one can get from satellites are incredibly complicated to reconstruct. The satellites in question, obviously, are nowhere near the troposphere. The troposphere is about 20km in thickness, and the satellites orbit at around 35,000km. What the satellites (or rather the MSUs onboard the satellites) measure, therefore, is the microwave radiation emitted by oxygen molecules in the atmosphere. That radiation bears a complex relationship to the temperature of air in which the molecules are found. The microwave radiation is measured in a variety of different "channels," each one of which bears a complicated relationship to the altitude where the oxygen molecules are found. The reconstruction of the temperature from the radiance, moreover, depends on many details of the methods used to do the calculation and on many auxiliary assumptions – especially the degree of decay of the sensors and the amount that the satellite in question has drifted off its initial orbit. This, again, is a terrific illustration of Suppes' idea of a *model of the experiment.* Getting a useful model of the data out of these satellites requires complex modeling that sorts out all of the complicated relationships between: the orbits of the satellites themselves and the ways in which those orbits have decayed over time; the connection between temperature and radiance from the various channels at various orbital positions; the relationship between altitude and the contribution to the radiance found in the various channels, etc. The UAH's model of the data represented one attempt to tease out all of the various influences and produce a simple-looking graph of troposphere data over time. The UAH's model showed no warming, and hence conflicted with the predictions of global climate simulation models.

True to Duhem's script, defenders of the simulation models criticized the dataset, and defenders of the dataset criticized the simulation models. Over time, two rival groups making data models for the satellite data (that is, two rivals of the UAH) emerged: the Remote Sensing Systems (RSS) group, and the University of Maryland (UMd) group. Both RSS and UMd produced temperature data that showed significantly more warming than the UAH dataset. Both of the new sets were much more compatible with the simulation models and showed greater agreement with the surface data collected by weather stations. Remember: all three of the datasets were based on exactly the same "raw data."

But Christy, Spencer, and the UAH had an ace up their sleeves: the radiosonde (or "weather balloon") data. The UAH group claimed that their satellite dataset was "independently confirmed" by the radiosonde data. The radiosonde data, in other words, also seemed to show that there was no warming in the tropical troposphere. In Christy and Spencer's minds, the radiosonde data were especially trustworthy because weather balloons measure the temperature of the troposphere "directly." They actually physically carry thermometers to the point of interest, rather than inferring from distant measurements, as satellites do.

As Lloyd points out, there were two problems with this way of thinking. The first problem was that radiosondes were widely regarded as unsuitable for recording long-term temperature trends. There are several reasons for this. Radiosondes were designed for weather forecasting, not for producing data that are intercomparable across large intervals of time and space. Frequent changes and improvements in the way the instrumentation function make comparisons across time into an apples-and-oranges affair (this problem is a close parallel of the buckets story). Radiosonde temperature measurements have to be adjusted for solar heating and time of day. Finally, the irregular spacing of radiosondes (especially over the oceans, where they rarely venture) makes it extremely difficult to extract meaningful spatial averages from them. Critics of the radiosonde data eventually found that making different and, they argued, equally plausible, assumptions about their spatial distribution could produce radically different data sets; including sets with temperature trends going in opposite directions.

The second problem was with the very claim that the radiosonde data provided "independent confirmation" of the UAH satellite data set. In fact, the UAH group had used the radiosonde data as a way of choosing between different possible methods of doing satellite intercalibration. Thus, it was highly misleading to suggest that the balloon data provided "independent confirmation" when in fact the UAH data set was always already laden with the assumption that the weather balloon data were correct.

The resolution of the controversy, which ultimately vindicated the simulation models and discredited the UAH critics, was a tour-de-force of complex experimental reasoning, involving a collaboration of experts in climate modeling, statistical analysis, and in the collection of data from a variety of sources, all led by Benjamin Santer (Santer et al. 2008). Constructing the right model of the experiment was a very complex and interdisciplinary affair. Among other things, the experts used other, genuinely independent data sets, including sea surface temperature, water vapor measurements, and tropopause heights, all of which should bear systematic relationships with tropospheric temperatures, to produce independent estimates of the data. All of those other data sets better supported the RSS dataset and were much more consistent with the warming trends predicted by the simulation models. They also made much more sense in the light of the highly trustworthy land surface temperatures. Eventually, the RSS group identified errors in the way UAH had accounted for the degree to which satellites had drifted out of their initial orbits.

In the end, the controversy was definitely settled in favor of the RSS and of the simulation models. All three groups (RSS, UMd, and UAH) now show a warming trend. A report of a study by the US Climate Change Science Program was produced in 2006. Lead authors included not only Santer but also John Christy himself. They concluded:

> Previously reported discrepancies between the amount of warming near the surface and higher in the atmosphere have been used to challenge the reliability of climate models and the reality of human-induced global warming ... This significant discrepancy no longer exists because errors in the satellite and radiosonde data have been identified and corrected. New data sets have also been developed that do not show such discrepancies."
> (Wigley et al. 2006)

Lloyd summarizes the attitude adopted by the Santer-led group as follows:

> Santer et al. claim that their work involves using a combination of observations, theory and models. They see the data as enmeshed with theory and models and their assumptions, and as constructed, rather than found and reflecting reality in a straightforward way ... As noted by van Fraassen, a complex empiricist approach "does not presuppose an impossible god-like view in which nature and theory and measurement practice are all accessed independently of each other and compared to see how they are related 'in reality' "(van Fraassen 2010). There is, in other words, no pristine separation of model and data. (Lloyd 2012)

This is the attitude to data, theories, and models that Lloyd calls "complex empiricism." "Direct empiricism," on the other hand, assumes that the weather balloons ought to be trusted more, since they "directly" measure temperature with thermometers that are physically located at the spatial point of interest.

## 2.5  Evidence of Warming, Continued

Another interesting observation from measured temperatures is revealed in Figure 2.2: warming across the globe has been far from uniform. Warming has been greatest in the Arctic, and least in the tropics. A cold winter in the United States, or a snowball brought into the Capitol building during a DC winter (Bump 2015) doesn't tell us much about global warming. But systematic trends in these phenomena, even regional ones, can sometimes be highly revealing. Some models suggest, for example, that the recent incidences of extreme winters in the Northeastern United States and Southeastern Canada are due to increased energy being channeled into the jet stream. Just as a car with more kinetic energy will tend to career more wildly down a slippery road, a jet stream with more energy will swing around more wildly, more often bringing the Arctic Polar Vortex down to more southern latitudes.

### 2)  Glacier Records

In addition to evidence in the form of recorded temperatures, there are also records of glaciers, which are of course most rationally compelling when they are presented in the form of numerical measurements, as in Figure 2.3, but are perhaps most rhetorically compelling when presented in images, as in Figure 2.4.

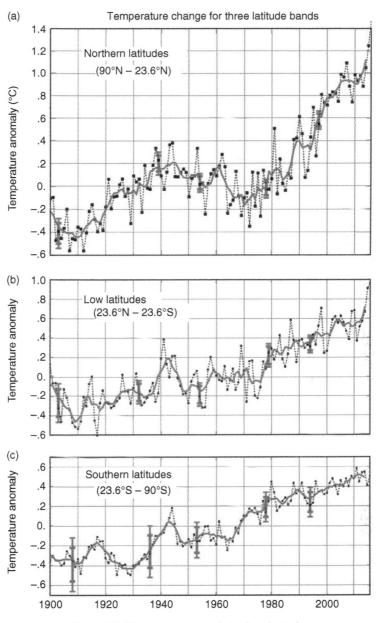

Figure 2.2 Temperature anomaly in three latitudes.
*Source*: adapted from http://cdiac.ornl.gov/trends/temp/hansen/
three_hemispheres_trend.html

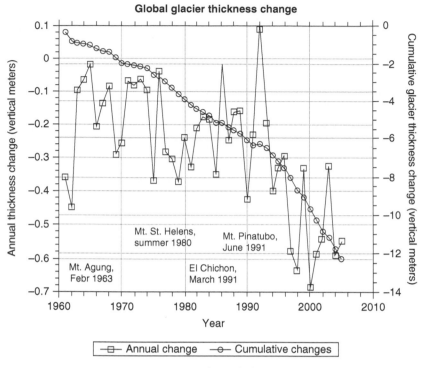

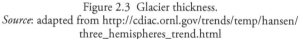

Figure 2.3 Glacier thickness.
*Source*: adapted from http://cdiac.ornl.gov/trends/temp/hansen/
three_hemispheres_trend.html

### 3) Arctic Sea Ice

The diagram and chart depicted in, respectively, Figures 2.5 and 2.6 reveal a third observational source of evidence for a warming trend over the last 50 years: the measured extent of Arctic sea ice. The chart also reveals something about our best models of climate: they leave open a fair degree of uncertainty regarding at least some of the details of the future pace and characteristics of climate change. None of our models predicted that the Arctic sea ice would retreat as fast as it has in the recent past. But this also shows that admissions of uncertainty regarding our models should not necessarily be equated with doubts about the severity of future outcomes. It is just as easy for our models to err on the side underestimating future dangers as the contrary. Sea ice also illustrates well the point about regional variation. You may have heard from skeptics that Antarctic sea ice has

Figure 2.4 A pair of northeast-looking photographs, both taken from the same location on the west shoreline of Muir Inlet, Glacier Bay National Park and Preserve, Alaska. They show the changes that have occurred to Muir Glacier during a period of more than a century, between the late-nineteenth century and August 11, 2005.
*Source*: USGS photograph by Bruce Molina

Figure 2.5 The 2016 Arctic sea ice summertime minimum, reached on September 10: 911,000 square miles below the average area of sea ice during the 1981–2010 period, which is shown with the line superimposed on the photograph.
*Source*: NASA Goddard's Scientific Visualization Studio/C. Starr

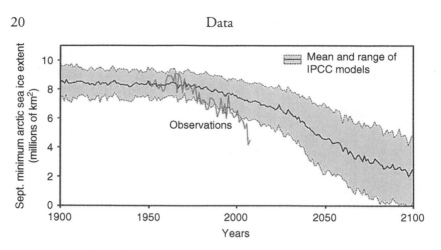

Figure 2.6 Graph of observed Arctic sea ice along with the mean and range
(one standard deviation) of forecasts by IPCC models from 1900 to 2100.
*Source*: adapted from Copenhagen diagnosis (Allison et al. 2009)

increased in the recent past. This is true, but the loss of Arctic ice vastly
swamps the gain in Antarctic ice.[3] What matters most, of course, is the net
change in sea ice around the globe. Climate scientists have explanations,
moreover, for why increases in Antarctic ice are consistent with a warming
trend. One explanation is that weather patterns near the Antarctic could
be pushing more cold air over the ocean, creating more sea ice. A second
is that the melting of the ice over Antarctic land sends large quantities of
fresh water, which freezes faster than salt water, into the ocean. Again, the
earth's climate is never simple.

### 4) Ice Sheets

An ice sheet is defined, in the earth sciences, as a mass of glacial land
ice more than 50,000 km$^2$ (or 20,000 mi$^2$). There are two such entities
on the planet: one that covers most of Greenland and one that covers
Antarctica. Together, they comprise more than 99 percent of the earth's
fresh water. If the Greenland ice sheet were to melt in its entirety, it would
raise global sea level by six meters. If the Antarctic ice sheet were to do the
same, it would raise sea level by 60 meters. The Greenland ice sheet has

---

[3] Indeed, in the time between when I wrote the first draft of this chapter and when this book went into
production, the 2016–2017 season brought a dramatic new global sea ice minimum, which included
a retreat in Antarctic ice.

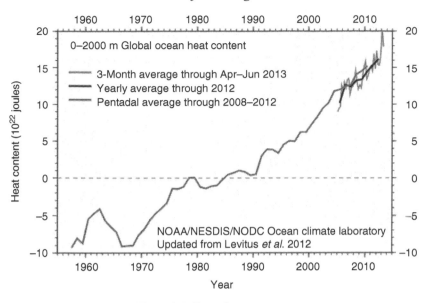

Figure 2.7  Ocean heat content.
*Source*: adapted from NOAA – www.nodc.noaa.gov/OC5/3M_HEAT_CONTENT)

experienced significant change, with summer melt having increased by 30 percent since 1979, though this has been partially offset by increased snowfall at higher elevations. Antarctica has not yet experienced as much warming as the rest of the planet but two areas of its ice sheet have been affected on the peninsula and on its western edge. The future of the ice sheets is one of the most controversial aspects of climate forecasting (in part because of the poor performance of models on this metric), but it is extremely important, for two reasons. First, the potential contribution to sea level rise would be enormous if any significant section were to collapse. Second, the positive feedback effects of ice sheet melting are large, since ice sheets reflect sunlight back into space and therefore have a cooling effect on the planet.

### 5)  Ocean Heat

Most of the additional heat trapped by the greenhouse effect is actually stored in the oceans, since water holds much more heat than the atmosphere. Figure 2.7 shows that ocean heat is increasing at a very steady rate.

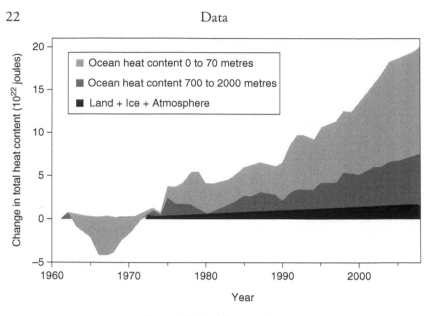

Figure 2.8 Total heat content.
*Source*: adapted from Nuccitelli et al. (2012)

### *6) Sea Level*

Changes in sea level are estimated from data from satellites and from tide gauges. Based on these data, global average sea levels are estimated to have risen 195mm from 1870 to 2004, and are estimated to have been rising between 2.6 and 2.9mm/year since 1993 (hence the rise has accelerated over the last decade). The principle causes of sea level rise are thermal expansion of the oceans (water expands as it gets hotter) and the melting of ice sheets and glaciers. Contrary to some popular beliefs, the melting of sea ice does not raise sea levels significantly since the buoyancy of ice is what is holding it above the surface in the first place, so when the ice contracts as it melts, this exactly offsets the volume of ice that plunges below the surface.

### *7) Temperature and Other Climate Proxies*

In addition to knowing about the climate of the last 150 years, climate scientists want to know about climate conditions that prevailed over the earth's history – the paleoclimate. This is useful both for getting a sense of the degree of internal variability of the climate system, and also for evaluating the skill of climate forecasting models. Because levels of $CO_2$ have varied in the past, it can also be helpful in estimating the relationship between $CO_2$ levels and temperature. To gain information about the climate prior

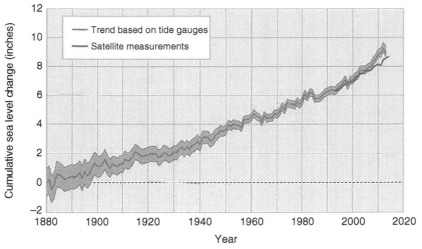

Figure 2.9  Sea level change.
*Source*: adapted from Nuccitelli et al. (2012)
Data sources:
- CSIRO (Commonwealth Scientific and Industrial Research Organisation). 2015 update to data originally published in: Church, J.A., and N.J. White. 2011. Sea-level rise from the late 19th to the early 21st century. *Surv. Geophys.* 32:585–602. www.cmar.csiro.au/sealevel/sl_data_cmar.html.
- NOAA (National Oceanic and Atmospheric Administration). 2015. Laboratory for Satellite Altimetry: Sea level rise. Accessed June 2015. http://ibis.grdl.noaa.gov/SAT/SeaLevelRise/LSA_SLR_timeseries_global.php.

For more information, visit U.S. EPA's "Climate Change Indicators in the United States" at www.epa.gov/climatechage/indicators.

to the late nineteenth century, scientists use what are usually called "climate proxies." The term highlights the fact that the variable in question is not being measured "directly." Of course, for reasons we noted above, this is somewhat misleading, since hardly anything of interest is ever measured "directly."

Typical climate proxies include ice cores, tree rings, pollen, boreholes, corals, lake and ocean sediments, speleothems (stalactites, stalagmites, and others), and fossils. Each kind of climate proxy has its own "span" and "resolution." "Span" refers to the period of time over which data from a particular kind of proxy is available. Tree rings can provide data from a few thousand years back, ice cores a few hundred thousand years, and fossils for millions of years. "Resolution" refers to how temporally fine-grained the data are. Tree ring data, for example, is annual. Ocean sediment data, on the other hand, gives values about once per century. Reconstructing the past climate from proxies, and estimating the relevant degree of uncertainty from these records, is an extremely complex science unto itself. Controversies frequently arise when different proxy sources conflict with one another or with "direct" data or with models. The history of paleoclimatology is a textbook case of Duhemian auxiliary hypothesis modification, as paleoclimatologists have frequently had to update their knowledge of the relationships that prevailed between climate variables and the marks they left on climate proxies in response to anomalous findings. What is clear, however, is that the degree of natural variability that we can observe over the last few thousand years has been eclipsed by changes over the last century.

## 2.6 Conclusion

I conclude the chapter with a simple moral: outside observers of climate controversies would be well served to keep in mind that the simplest, most straightforward, and most commonsensical arguments regarding the relations between data and hypotheses in climate science are not always the best-considered ones. The simple, the straightforward, and the commonsensical can be the enemies of the truth in a messy world. Much simple-minded criticism of climate science depends on the assumption, to the contrary, that the simple, the straightforward, and the commonsensical are always good guides to the truth. As in the famous case of "climategate" (see Box 2.3), critics call foul when they see raw data being transformed into data models with complex chains of reasoning. They make heavy use of the tactic of calling nefarious any attempt on the part of climate scientists to use "tricks" to transform raw data into data models, data sets, or "phenomena." The reality is that lay people are rarely well situated to appraise how well various data sets are supported by raw data. In the real world of climate science, as opposed to the fantasy world of critics, connecting the complex chain of dots between raw data and hypothesis can require an enormous degree of scientific sophistication and the cooperation of a variety of different kinds of experts.

## Suggestions for Further Reading

Dessler, Andrew. 2016. *Introduction to Modern Climate Change*. Excellent introduction to the science and policy of climate change. Very introductory; accessible to students with little science background.

IPCC, AR5, Synthesis Report, Introduction and Topic 1: "Observed Changes and their Causes." A concise summary of the observational evidence for a changing climate, the impacts caused by this change and the evidence of human contributions to it. No document better represents the consensus of the scientific community on these matters.

---

### Box 2.3 Climategate

"Climategate" is a common term used in the media and in the blogosphere to label a "controversy" that was manufactured by climate change deniers following the November, 2009 release of illegally obtained emails from the climate research group at the University of East Anglia, UK. Numerous subsequent investigations found no evidence of fraud or manipulation of data, but skeptics were undeterred. In a nutshell, critics of climate science were exploiting the Duhem problem to make it look like climate scientists were fabricating their evidence, putting particularly strong weight on an email written by Phil Jones that wrote: "I've just completed Mike's *Nature* trick of adding in the real temps to each series for the last 20 years (i.e. from 1981 onwards) and from 1961 for Keith's to hide the decline." The "trick" was merely the well-established technique of using tree rings to plot temperature over several centuries, but to transition to instrument readings starting in 1960 – the point at which they are considered to be more reliable. (And in fact this is almost always carefully labeled in any chart showing these data.) RealClimate, an important climate change blog, summarized the episode quite pithily as follows: "More interesting is *what is not contained* in the emails. There is no evidence of any worldwide conspiracy, no mention of George Soros nefariously funding climate research, no grand plan to 'get rid of the MWP,' no admission that global warming is a hoax, no evidence of the falsifying of data, and no 'marching orders' from our socialist/communist/vegetarian overlords. The truly paranoid will put this down to the hackers also being in on the plot though."[4]

---

Suppes, Patrick. 1969. "Models of Data." Seminal work in the philosophy of science on the relationship between theories, models, and data.

Bogen, James and James Woodward. 1988. "Saving the Phenomena." Another seminal work very much in the spirit of "Models of Data." The authors argue that the relevant notion of a phenomenon in science should be understood as a highly crafted piece of scientific work, not as something that is directly given by the world.

Lloyd, Elisabeth A. 2012. "The Role of 'Complex' Empiricism in the Debates about Satellite Data and Climate Models." Excellent account of the satellites controversy and its implications for our understanding of both climate science and the nature of scientific data.

Realclimate.org. A blog run by climate scientists that designed to keep journalists and the interested public informed. They "aim to provide a quick response to developing stories and provide the context sometimes missing in mainstream commentary." Posts by Gavin Schmidt are likely to be especially interesting to philosophers.

SkepticalScience.com. A website that contains good discussions of standard arguments by climate change deniers and useful summaries of current climate research.

---

[4] (RealClimateGroup 2009)

# 3

# *Models*

## 3.1 Introduction

In the last chapter, we looked at a sample of the variety of data that support the claim that the climate has changed in the last 150 years. But the simple data themselves are relatively silent about causes. Where does the idea come from that such change might be attributable to human activities – in particular to the release of carbon dioxide into the atmosphere? The idea actually goes back to 1896, when the Swedish scientist Svante Arrhenius calculated the atmospheric cooling and heating effects of raising or lowering the concentrations of carbon dioxide in the earth's atmosphere. It was not until 1938, however, that the British engineer Guy Callendar gathered evidence that both the temperature, and the levels of carbon dioxide in the atmosphere, had been rising for several decades. He then presented new evidence based on spectroscopy that $CO_2$ was a heat-trapping gas in the atmosphere. So what does a simple calculation, like the one Arrhenius could have performed, look like?

## 3.2 Energy Balance Models

To begin, we'll look at what's called a zero-dimensional energy balance model of the earth. It's called "zero-dimensional" because it does not allow for any spatial variation in any variables, but really it treats the earth as the surface of a sphere, as it would look to you if you were looking at it from the surface of the sun. It is called an energy balance model because it asks what happens to the earth's temperature if and when it achieves a balance between incoming and outgoing energy – so it does not allow for any variation in time for any variable either.

The only sources of incoming and outgoing energy are radiation, which we can measure in Watts per square meter ($Wm^{-2}$). So we have

Incoming radiative energy $E_{in}$ and
Outgoing radiative energy $E_{out}$.
And since we want an equilibrium model, we set $E_{in} = E_{out}$.

Let's call the energy per square meter that the sun delivers to the earth the incident solar radiation (or sometimes "insolation"), $S_0$.

Since the earth presents a disk-shaped face to the sun, it has area $\pi R^2$, so it receives $\pi R^2 S_0$ in incoming radiation.

Let's also assume that some fraction $\alpha$ of that (which we call the albedo) is reflected back into space and that the remainder $(1-\alpha)$ (the "co-albedo") is absorbed. We now have a formula for $E_{in}$

$$E_{in} = (1 - \alpha) \pi R^2 S_0 \tag{3.1}$$

What about outgoing energy? Here we will treat the earth as a spherical black body that obeys the Stefan-Boltzmann law, which says that a black body radiates away heat in proportion to the fourth power of the temperature (T, in degrees Kelvin), with the constant of proportionality, $\sigma$, called the Stefan-Boltzmann constant. This gives us a formula for $E_{out}$

$$E_{out} = 4\pi R^2 \sigma T \quad (4\pi R^2 \text{ being the surface area of a sphere}) \tag{3.2}$$

If we set $E_{in} = E_{out}$ and then solve for T we get

$$T = \left( \frac{(1 - \alpha) S_0}{4\sigma} \right)^{\frac{1}{4}} \tag{3.3}$$

$$S_0 = 1,368 \ Wm^{-2}$$

$$\sigma = 5.67 \times 10^{-8} \ Wm^{-2} K^{-4}$$

This gives us that T = 254.8K or –18.3°C

We can call that the "effective temperature." But the actual temperature of earth is more like 15°C.

So we are wildly off! How can we improve our model? In the above, we made what turns out to be a significant idealization. We assumed that the radiation was unobstructed on the way in and on the way out. It turns out that the first idealization was reasonable enough, but the second one wasn't. That's because the incoming radiation is very short wavelength and the outgoing radiation is very long wavelength. The earth's atmosphere lets almost all of the former pass through, but the behavior of the long wavelength radiation is more complex. The atmosphere absorbs much of the energy of this radiation, and, like the earth, radiates that energy away. The energy is radiated away in all directions, including back toward the surface of the earth. Thus the incoming radiation consists of two components: the radiation received from the sun plus the energy reradiated back to the earth from the atmosphere. This is the famous "Greenhouse Effect." The effect is incredibly strong on Venus, which has an effective temperature of "only" –42°C, but an actual surface temperature of 462°C. To get the earth's actual temperature, rather than its effective temperature, we could complicate (or "de-idealize") our model in one of two ways. The simpler one would be to add a simple "emissivity coefficient," and call it $\epsilon$. The emissivity coefficient is just defined as the proportion of the energy radiated out by a body relative to what a black body would radiate. There are various ways of measuring the emissivity of the earth, and they generally settle, for the present time, at around $\epsilon = 0.62$.

A more complex method would be to try to model the physical causes of the earth's particular emissivity. But this would require us to model several different physical effects, including the spectrum of the radiation, the effects of the chemical contents of the atmosphere on the distribution of clouds and on other features of the atmosphere, and the physical topography of the earth on the creation and passage of each band of radiation.

For now, let's go the simpler route, and modify equation 3.2 to include only proportion $\epsilon$ of the outgoing radiation.

$$E_{out} = 4\pi R^2 \epsilon \, \sigma T \tag{3.4}$$

and now we get

$$T = \left( \frac{(1-\alpha)S_0}{4\epsilon\,\sigma} \right)^{\frac{1}{4}} \tag{3.5}$$

With $\epsilon = 0.62$, we get a much more realistic answer of T = 287.7K or 14.6°C.

What does this have to do with greenhouse gases? In our model, epsilon is just a constant. In the world, the amount of escaping radiation trapped by the atmosphere depends on a large variety of factors. But we do know with near certainty that, *ceteris paribus* (see Box 3.1), the presence of certain gases in the atmosphere will lower the emissivity. We know it from the field of spectroscopy, of which Arrhenius was an expert. It happens because gases are made out of molecules that vibrate in various ways, each of which corresponds to a particular wavelength of radiation (and because of the Doppler effect, those nearby) that they are keen to absorb. So each

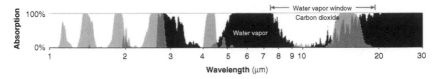

Figure 3.1 This graph shows the spectra where carbon dioxide and water vapor absorb radiation. Notice that water vapor leaves open a "window" that can be partially closed by carbon dioxide. Given its abundance and the rate at which human activities produce it, this makes carbon dioxide particularly important. So, it is pretty clear that the more carbon dioxide that is in the atmosphere, the lower $\epsilon$ gets and the higher is the temperature predicted by this model.

chemical component of the atmosphere traps different wavelengths of radiation in different proportions to the former's abundance. On our planet, it turns out that carbon dioxide is a potent agent for lowering emissivity. Given their current abundances, the actual components of our atmosphere that do the most trapping of energy are water vapor, carbon dioxide, methane, and ozone. By far the strongest of these is water vapor. But the amount of water vapor in the atmosphere is really a function of its heat (heat makes water evaporate), rather than the other way around, so it's not a potent independent variable. But Figure 3.1 illustrates why the impact of water vapor on emissivity makes carbon dioxide so important:

What does this zero-dimensional energy balance model, and this insight into the effects of $CO_2$ on emissivity, tell us about the likely cause of the recent warming pattern evident in our best models of the data? And what does it tell us about the likely effect of adding more carbon dioxide into the atmosphere?

## 3.3 The Nature of Models: Mediating Models

It is helpful here to take a minute to think about the nature of models like this, and consider what they are good for. The idea of a "model" is ubiquitous in the sciences, so much so that it is hard to resist the temptation to think the term is a catchphrase for a grab bag of different sorts of entities and activities. In Chapter 2, we talked about *models of data*. Closely connected to models of data are the *models of instruments* and *models of experiments* that inform our models of data. When we reconstructed the central tendency line and the error bars in Figure 2.1, we had to have a model of how our satellites behave and how the radiances they collect are related to the temperature of the troposphere. We also use the word "model" in, for example, the "standard model" in particle physics, or the "Lambda-CDM model" of the cosmos. In those cases, we are using the word "model" to signal that quantum field theory, in the first case, and the Big Bang theory of cosmology, in the second, are just theoretical frameworks that need to be fleshed out with the sorts of details and parameters that the "standard model" and the "Lambda-CDM model" provide. This is the notion of model that is best thought of as a *model of a theory*.

None of those, however, are exactly what we are interested in here. What we are interested in are models that are used to characterize a target system or phenomenon in such a way that various salient bits of theory and other mathematical regularities can be applied to draw out inferences about the target by reasoning with the model. It has become popular in recent philosophy of science to call these "idealized models," but I prefer to call them by an older name, *mediating models* (Morgan and Morrison 1999), for two reasons. The first reason is that many of the *other* kinds of models I mentioned above engage in idealization, so idealization is not unique to this kind of modeling. The second reason is that while it's true that all mediating models involve idealization, I don't think the *purpose* of these models is to idealize. The purpose, as I said above, is to be able to apply bits of theory, law or mathematical regularity to a target system – that is, to mediate between theory (and model-building principles with very sound epistemic credentials) and the world. Idealization is usually only a means to that end, even if it is almost always a crucial and profoundly important element of it.

It is also popular in philosophy of science to divide so-called idealized models into "Aristotelian idealization" and "Galilean idealization." On this account, the former consists of

> "stripping away", in our imagination, all properties from a concrete object that we believe are not relevant to the problem at hand. This allows us to focus on a limited set of properties in isolation. An example is a classical mechanics model of the planetary system, describing the planets as objects only having shape and mass, disregarding all other properties. (Frigg and Hartmann 2012)

Galilean idealizations, on the other hand, are

> ones that involve deliberate distortions. Physicists build models consisting of point masses moving on frictionless planes, economists assume that agents are omniscient, biologists study isolated populations, and so on. It was characteristic of Galileo's approach to science to use simplifications of

this sort whenever a situation was too complicated to tackle. (Frigg and Hartmann 2012)

This makes it seem like there are two different kinds of models and their functions are different.

Maybe that's true in some other domains, but I'm going to resist this divide because I think these kinds of activities overlap far too much in climate modeling. Our example highlights this. A zero-dimensional energy balance model involves deliberate distortions in order to tractably apply some of the theory it applies. But it also strips away properties that are not relevant to the problem at hand, so long as we properly understand what that problem is – which is to get the effect of one very important factor in isolation. (More on this later.) Indeed, the two generally go hand-in-hand for good reason: so-called Galilean idealization usually only works insofar as some degree of "Aristotelian" reasoning is in play. We will see this in a minute with our example.

So let's re-examine our notion of a mediating model with the example of a zero-dimensional energy balance model. The first thing we said was that a mediating model is "used to characterize a target system or phenomenon in such a way that various salient bits of theory and other mathematical regularities can be applied." That's what we did. We characterized the earth as a homogeneous black body being bathed in a uniform field of radiation with a uniform degree of reflectivity and a single-parameter greenhouse coefficient, etc. We did this so that we could apply bits of theory like the Stefan-Boltzmann law, the conservation of energy, and our little side-model that told us about the effect of carbon dioxide and water vapor on the passage of long-wavelength radiation. We used the model to mediate between those abstract bits of theory and our target system: the sun, the earth, and its atmosphere, etc. To be sure, we made massive idealizations. Did we do that in order to make the application of those bits of theory tractable, or because it allowed us to focus on a limited set of properties in isolation? Well, both. Our real goal was neither of those. Our real goal was getting a grip on the problem with well-credentialed theoretical principles we trusted (like the Stefan-Boltzmann law). To do that, we have to do some Galilean idealization. But unless we also understand our idealizations in the context of the understanding we achieved through Aristotelian idealization, we are bound to go badly wrong with understanding the limits of its usefulness.

What's the point of these (energy balance) models, after all, with all of their crazy distortions? Arguably, the point is to get at a basic, but obviously *ceteris paribus*, causal effect: long-wavelength-radiation-trapping

---

**Box 3.1  Ceteris Paribus**

*Ceteris Paribus* is a Latin phrase that means "other things being equal." In philosophy of science it usually means "other (possibly interfering) things being absent." When we say that some relation obtains *ceteris paribus*, we mean that we can expect it to hold only under certain expected conditions, and that it might fail if other intervening factors are present. In fact, we can mean two different things by this. We might mean that the relation actually does obtain when those intervening factors actually are absent (in other words that the intervening factors do, in reality, tend to be absent in relevant circumstances). Or we might only mean that, hypothetically, the relation would obtain if the intervening factors were absent, without any commitment whatsoever that the factors ever actually are absent. In our energy balance model, for example, we could conclude that carbon dioxide raises temperature *ceteris paribus* without being committed to the claim that the intervening factors that would push that effect around actually are ever absent in the real world.

---

molecules in the gases of the atmosphere can, *ceteris paribus* (more on this in a bit), make a planet hotter. But also arguably, the point is to be able to write down an analytically tractable model that can be used to make calculations without a supercomputer. The second works only insofar as the first is in place, and vice versa. This Galilean idealization will be disastrous if we are using the model to predict tomorrow's weather, or even the climate at the end of the century. They work because our goals for them are more modest and because we keep our goals modest – to get at a simple, *ceteris paribus* causal effect – to get at a basic mechanism rather than make accurate predictions – we can do it by writing down a highly simple and tractable model.

There has also been a lot of ink spilled in philosophy of science about how models represent their targets. Do they have to be isomorphic to their targets? Partially isomorphic? Homomorphic? Similar? If the latter, then similar in what respect? Is there a metric of similarity? Etc. All of these are ways of specifying what it is for a model to represent its target – for there to be a *fit* between model and target. This is important insofar as one thinks of models principally as vehicles of representation, rather than as devices for making various kinds of inferences. For thinking about climate science and climate models, though, I think it is more helpful to think of models as being adequate for purpose or not, rather than qualifying as fitting their targets or not (Parker 2010b). The kind of question that will most often interest us is, for some particular purpose or set of purposes, to some desired degree of accuracy, and across some domain of targets, can the model be a reliable tool for reasoning about the target in a way that meets those purposes? For lots of models and lots of purposes, deciding that the model is adequate for purpose, and making a convincing case to others that the model is adequate for purpose, can be quite complicated – and the criteria that will have to be met in order to provide that assurance might be extremely hard to pin down. If we set aside the epistemological question of how we *decide* when a model is a good one, and ask instead what *it is* for a climate model to be a good one, I think we can offer a fairly straightforward answer, and it is this: to be a good model is *purpose relative*. It is simply to be adequate to the various tasks to which we intend to put the model, whatever those are. It might be for making predictions of a certain kind, or for uncovering causal relationships, or to convey something pedagogically, or to gain understanding, or one or more of many other things. It is also usually *relative to a standard of accuracy*. That is, whether a model counts as adequate for such a purpose might depend on how accurately we hope to meet that purpose: particularly if the purpose is prediction. And finally, it is *domain relative*. A model might be adequate to purpose in one domain, but brittle when it leaves that domain.

## 3.4 Conclusion: How Good Are Energy Balance Models?

With all of that preamble out of the way: how good are the zero-dimensional energy balance models of equation 3.3? Clearly, it depends on purpose, desired degree of accuracy, and domain. If our purpose is to predict the temperature of planets, then we had better either set our degree of accuracy very low, or restrict our domain of interest to planets without atmospheres, if we hope for the model to plausibly be adequate for purpose. We need only look at Venus to see that. Equation 3.3 gets Venus wrong by almost 500 degrees! But if the purpose of the model is to help build a basis for reasoning about the primary determinants of planet temperatures, then it turns out to be a fairly useful model, because it helped us to infer that emissivity is an important component.

If we move to the model behind equation 3.5, the predictive quality of the model has improved, in that its accuracy can be quite good for a broader domain of planets, but only if we have a reasonable determination of the emissivity from somewhere outside the model. The model we wrote down is useless for *predicting* that emissivity if we don't already have it. It is also useless for predicting something which is pretty important to understanding our own situation: if I change one of the variables in the equation, say the incident solar radiation, or the emissivity, or the albedo, how long does it take the planet to newly reach equilibrium? What path does it follow? The model is useless for these questions because it has no dynamics. It only tells us, at best, where the system will end up at the end of time.

The dynamics of the atmosphere, of the oceans, and of other features of the climate system are highly non-linear and complex, and so dynamical models of the climate are almost necessarily *computer simulations* of the climate. We will have a lot more to say about computer simulations of climate, and their epistemology, in the rest of the book. For now it is worth pausing and asking how much we can know without these tools. How much do we know just from data and observations, and from simpler models like energy balance models? In particular, how well can we answer two related questions without the help of simulations?

1) How much of the observable warming trend of the last century of so can be attributed to human activities?
2) What is the climate sensitivity to carbon dioxide?

How well we can answer these two questions using only "basic science" (no simulation) is a matter of both some complexity and some controversy. The answer is complex, in part, because energy balance models and evidence from data and proxy data are most helpful in answering the two questions when they are used in conjunction with knowledge we get from simulations. This should not be overlooked or underestimated. Everything we believe about the climate is part of a complex web of beliefs that are mutually intersupporting, and even what might sound like common-sense, back-of-the-envelope reasoning from simple models and data might in fact be supported by tacit knowledge acquired with the help of simulation models. (It is possible, for example, to write down simple analytically solvable dynamical models of the climate, but they often need parameter inputs that can only come from simulations.) It is very hard to pinpoint the exact sources of the confidence we have in individual claims. There are, nevertheless, a few things we can say.

We'll start with the second question. The question concerns the numerical relationship between the concentration of carbon dioxide in the atmosphere and the temperature of the planet. But we should define the term "climate sensitivity" more carefully. There are two kinds of climate sensitivity, and each can be defined in two ways. The first kind is "equilibrium climate sensitivity." This is most often defined as the amount which a sustained doubling in the quantity of $CO_2$ in the atmosphere would raise the equilibrium global surface temperature. It is also sometimes defined more abstractly as the change in global surface temperature we would expect, at equilibrium, from a unit amount (in W/m²) of "radiative forcing." Radiative forcing is the net change in the earth's radiative budget at the troposphere caused by a perturbation to the system; where the perturbation could be anything from a change in the amount of "insolation." (This insolation could come from the composition of the atmosphere, the albedo, or anything else that affects the energy balance.) We can define sensitivity either of these two ways and think that we are after the same thing because it is generally agreed that, at least to first order, the sensitivity to radiative forcing is independent of the cause of the forcing. In addition to equilibrium sensitivity, we are also sometimes interested in the "transient climate response," which can be defined in a variety of ways but which tries to get at what a gradual doubling of $CO_2$ would do over a period of about 20 or 30 years.

Regarding equilibrium sensitivity, we can use simple models to calculate that a doubling of $CO_2$ corresponds to a forcing of about 3.7W/m². The equilibrium response to this degree of forcing, *ceteris paribus*, can be easily calculated from equation 3.5, and it corresponds to about 1⁰C of warming. This much follows from simple modeling and no one disputes it. But the *ceteris paribus* clause here is extremely important, and it does most of the heavy lifting, because this calculation assumes that the rise in temperature caused by the forcing will not itself bring about changes to the system that will *contribute forcings of their own*. That is, this calculation assumes there are no feedbacks! In reality, we know there are at least four very significant sources of feedback: water vapor; clouds; ice-albedo; and the lapse rate (the *rate* at which atmospheric temperature decreases with increase in altitude). Each one of these four variables changes in response to a changing temperature. When they do, they cause a forcing of their own which can then go back and affect the temperature. Feedback! Of the four, only the water vapor feedback can even be estimated very well with a simple non-dynamical model (if we assume relative humidity stays the same as the temperature rises) and that calculation tells us that the climate sensitivity is approximately doubled by this mechanism. The rest depend significantly on the dynamics and internal variability of the components of the climate system. Clouds, for example, both lower emissivity (by absorbing outgoing radiation) and add to albedo (by reflecting back sunlight). So they are both heating and cooling – with lower clouds probably being net heaters and higher clouds net coolers. The lapse rate also affects emissivity, as each layer has its own temperature and hence radiates at different rates. (Simulations tell us that the lapse rate is a source of negative feedback.) Ice and vegetation (which depends in complex ways on many features of the climate system) can also affect albedo. All these things are products of the complex dynamics of the atmosphere, so no non-dynamical model is going to get them right – they are above the pay grade of an energy balance model. And no energy balance model is going to tell us anything at all about the transient climate response. Or at least: no energy balance model that does not use inputs from a dynamical model is going to be able to do this.

Climate sensitivity, and even the transient response, can also be estimated from various sources of data. The best of these come from ice core proxy data of glacial and interglacial periods, from instrument data during volcanic eruptions, and from the overall instrument record of the last 150 years. All these sources have problems and they work best when they work in conjunction with auxiliary inputs and models that are in turn supported by simulation models. The only thing that is almost completely uncontroversial here is that the ice core proxy data from interglacial periods probably fix an upper limit on the value of equilibrium climate sensitivity at around 6⁰C.

What about the first question? What do we know from energy balance models and observations alone that is relevant to one of our central questions: can the observed change in our climate over the last 150 years be predominantly attributed to human activities? How strong is the argument for anthropogenesis that is independent of dynamical models? This is also a matter of some controversy, but the IPCC takes a rather pessimistic view.

According to the IPCC, detecting a change in the climate, based on observations, requires determining that "the likelihood of occurrence by chance due to internal variability alone … is small." This, in turn, requires an estimate of internal variability, generally derived from a "physically based model." Furthermore, going on to attribute the detected change to a specific cause (such as human activity) typically involves showing that the observations are, "consistent with results from a process-based model that includes the causal factor in question, and inconsistent with an alternate, otherwise identical, model that excludes this factor." Indeed the authors of the IPCC report are quite clear that, "attribution is impossible without a [process-based] model." "We cannot observe a world in which either anthropogenic or natural forcings are absent, so some kind of model is needed to set up and evaluate quantitative hypotheses: to provide estimates of how we would expect such a world to behave and to respond to anthropogenic and natural forcings." (IPCC, AR5, WG1, 10.2.1)

On the other hand, we can reason our way to the claim that, *ceteris paribus,* adding gasses like carbon dioxide and methane, gasses that we know trap radiation at wavelengths through which much of our planet's emissivity takes place, will decrease the emissivity, and make the planet hotter. Relatively simple reasoning suggests that feedbacks are much more likely to be net positive than negative. We have observational evidence that the planet's carbon dioxide content has gone up and that it has gotten hotter to a degree that looks, by all accounts, to have exceeded the natural variability of the last several thousand years. If we put these three things together, some would say that we get a pretty strong prima facie case for anthropogenesis. In any case, it is fairly clear that the case for anthropogenesis (and other hypotheses about the climate) are at least bolstered by dynamical models – computer simulation models. So we will turn to computer simulations of the climate in the next chapter.

## Suggestions for Further Reading

Morgan, Mary and Margaret Morrison. 1999. *Models as Mediators.* An influential collection of essays on the role of models in science that emphasizes particularly well the representational and inferential role of models that I have in mind with my own use of the expression "mediating model."

Frigg, Roman and Stephan Hartmann. 2012. "Models in Science," *The Stanford Encyclopedia of Philosophy.* A concise summary of all the work in philosophy of science on scientific modeling from the last 40 years. Contains an extensive bibliography.

Parker, Wendy. 2009. "Confirmation and Adequacy-for-Purpose in Climate Modelling." A vigorous defense of the claim, which we took aboard in this chapter, that climate models should be confirmed as "adequate for purpose" and not truth or overall fitness.

Suárez, Mauricio. 2004. "An Inferential Conception of Scientific Representation." Argues for a picture of scientific modeling according to which the fundamental feature of models is that they facilitate inferences about their targets.

4

# *Simulations*

## 4.1 Introduction

In the last chapter, we encountered the claim that any model that hoped to capture the internal dynamics of the climate – and that could therefore capture things like feedbacks and transient responses – would have to be a computer simulation model. Atmospheric scientists learned this the hard way in the early part of the twentieth century. They tried to use paper and pencil calculations to predict and describe the large-scale motions of the atmosphere – with zero success. It wasn't until the early 1950s that meteorologists were able, using the newly built digital computers, to produce crude predictions of a day's worth of the basic features of the atmosphere. It wasn't much; but it was enough to push on. By the mid to late 1960s, computer predictions of the next day of weather began to do better than the educated guesses of trained human forecasters (Weart 2010, p. 210).

Why do we need computer simulations to predict even the coarsest and most short-lived features of the atmosphere? The core behavior of the atmosphere can be modeled with three simple laws: Newton's laws of motion as they apply to parcels of fluid, the conservation of mass, and a simple thermodynamic equation that allows us to calculate the heating effect on each parcel of air via a parameterized value of the radiation from the sun. Unfortunately, what we get out of this is a coupled set of non-linear partial differential equations for which we have no closed form solution. We can at best hope to get a *numerical approximation* of how a system governed by such equations should behave. We do this by trying to transform the original (continuous) differential equations into discrete difference equations that approximate them, and we use the computer to solve the latter step-by-step over discrete intervals of time for discrete points in space. Rather than a function that tells us values for variables like temperature and pressure for arbitrary points in time and space, the computer provides numerical values for these variables on a space-time grid.

Modern climate models of the most advanced kind do much more than model just the circulation of the atmosphere. The atmosphere, after all, is only one part of the climate system – which consists of not only the atmosphere but also the hydrosphere (including the rivers and seas), the cryosphere (ice caps and sheets), the land surfaces, and the biosphere, and all the complex interactions between them. Not only does a climate model need to couple the circulation of the atmosphere to the circulation of the oceans, but the atmospheric component must include representations of physical features like clouds, precipitation, and aerosols; the ocean component must include sea ice dynamics, iceberg transport of fresh water, currents, and wave dynamics; the land component will include precipitation and evaporation, streams, lakes, rivers, etc.; and the ice sheet component will include thickening and thinning and cracks and fissures.[1] A full Earth System Model (ESM) also tracks sources and sinks of carbon into and out of the biosphere and other systems. State-of-the-art climate models, in other words, are massively complex, and a full treatment of everything they do and how they do it is far beyond the capacities of a text like this one. For philosophers of science interested in climate modeling, there is a massive amount of knowledge, lore, and expertise to try to absorb.

In what follows I'll try to focus on four relatively modest goals. In section 4.2, we'll begin by exploring a few features of computer simulations in general. In section 4.3, we will discuss the work that is done with climate simulations – the purposes to which they are put. In section 4.4 we will first explore some of the technical features of climate simulations: the equations that underlie them, their computational realizations, and the physical and biological processes they try to incorporate. Then we will turn to some of the features of climate simulations that are of explicitly epistemological significance: their degree of modularity, their reliance on parameterization, and the ways in which they are tuned to fit existing data (Winsberg 2013).

## 4.2 What Is Computer Simulation?

It is useful to understand the expression "computer simulation" in two different ways. The first is as the name for the execution of a particular sort

---

[1] See www.gfdl.noaa.gov/earth-system-model and also Box 4.4 for a description of one of the "flagship" American models.

of computer program (a "computer simulation model") – one that explores the approximate behavior of a mathematical model, for a particular value of initial and boundary conditions, using step-by-step methods. The second is as a name for a comprehensive set of methods for studying certain sorts of complex system. Let's call the first one "the narrow notion" and the second one "the broad notion."

On the narrow notion of a computer simulation, one run of the relevant program on the computer is one computer simulation of the system. The algorithm instantiated by the program takes as its input a specification of the system's state (the value of all of its variables at all spatial locations) at some time, t. It then calculates the system's state at time t+1. From the values characterizing that second state, it then calculates the system's state at time t+2, and so on. When run on a computer, the algorithm thus produces a numerical picture of the evolution of the system's state, as it is conceptualized in the model. It produces, in other words, a value for each variable at each spatial location (if the model is indeed spatially discretized) and for each moment in time. This sequence of values for the model variables can be saved as a large collection of "data" and is often viewed on a computer screen using methods of visualization. Often, but certainly not always, the methods of visualization are designed to *mimic* the output of some scientific instrument – so that the simulation appears to be measuring a system of interest.

In the kinds of simulations that interest us, simulations of the climate, step-by-step methods of computer simulation are used because the model of interest contains continuous (differential) equations (which specify continuous rates of change in time) that cannot be solved analytically. Creating the step-by-step algorithm involves "discretizing" the relevant equations (the turning of equations that describe continuous rates of change into discrete equations that describe how they change over discrete intervals of time). Discretization can be carried out in innumerably many different ways – all with their own mathematical properties – but the simplest, conceptually, is by carving the world of the simulation up into a regular grid and assigning variable values to regular intervals of space and time. Other more sophisticated implementations of discretization exist (spectral methods, adaptive grids, mass discretization, etc.) but the particulars of the various implementations won't affect many of the issues we discuss. It should be emphasized though that, although it is common to speak of simulations "solving" those equations, a discretization can at best only find something which approximates the solution of continuous equations, to some desired degree of accuracy. We should also, when speaking of "a computer simulation" in the narrowest sense, strictly speaking, have in mind a particular implementation of the algorithm, with a particular discretization scheme, on a particular digital computer, written in a particular language, using a particular compiler, etc. There are cases in which different results can be obtained as a result of variations in any of these particulars, and if we really want to individuate simulation by runs of a program, we should individuate them in terms of all the details that can make a difference.

More broadly, we can think of computer simulation as a comprehensive method for studying systems. In this broader sense of the term, it refers to an entire process. This process includes choosing a model; choosing the best way of implementing that model in a form that can be run on a computer; calculating the output of the algorithm; and visualizing and studying the resultant data. The method includes this entire process – used to make inferences about the target system that one tries to model – as well as the procedures used to sanction those inferences.

## 4.3 The Purposes of Climate Modeling

Computer simulation can be used for a wide variety of purposes: to gain understanding, to predict, to develop designs, and to educate, to give at least a partial list. In the last chapter, we emphasized that, at least in the context of climate science, models are best evaluated vis-à-vis their adequacy-for-purpose. This is especially true with regard to simulation models of climate, so it is worth saying a few preliminary things about the purposes to which climate simulations are *generally* put. Though we of course use climate models to gain understanding of the climate, and for other purposes, their most important, and also most epistemically fraught, application is for prediction. But it should be emphasized that they are used for *climate* prediction[2], not for weather prediction, and these are qualitatively different sorts of activities. Climate simulation models are close cousins to weather simulation models. Historically, they were partly fashioned out of them. But the nature of the tasks to which the two are put divides them quite profoundly. Weather models are used for making spatially and temporally fine-grained predictions of, well, the weather. We use weather models to make predictions of the form: given that the best summary of the present atmospheric conditions we have is such

---

[2] Or, to really emphasize the difference, what the IPCC prefers to call "climate *projection*." But in the more ordinary parlance of science and modeling, this is still a kind of prediction.

and such, what does our model tell us are likely to be the atmospheric conditions next Tuesday afternoon in Tampa, Florida? But as every viewer of the Weather Channel knows, weather prediction degrades in quality very quickly after about five days.

So what good are climate models given that climate is a concept that is tied to periods of time of at least 30 years or so? The answer lies in the fact that climate variables are, generally speaking, *statistical summaries*, and indeed usually very coarse-grained summaries, either in the form of averages or degrees of variation of weather variables. They are coarse-grained in the sense of involving, usually, at least 30-year averages, and also in the sense of involving either global averages or at least averages over spatial regions that are of roughly continental scale. But also, climate models are not generally in the business of predicting climate variables *conditional on present conditions*. In a sense, the idea of predicting the *climate* conditional on the present conditions is a bit of an oxymoron. The cliché aphorism, remember, is that climate is what you expect, and weather is what you get. The notion of "what you expect" is at least supposed to be independent of what you *happen to have gotten* today. To the extent that the present weather conditions contribute to the future evolution of the climate system, they contribute to aspects of its *internal variability*, which is precisely what we are trying our best to leave out when we talk about climate. Quoting the IPCC: climate models are used to make "Climate projections": "climate change experiments with models that do not depend on initial conditions but on the history and projection of climate forcings" (IPCC, AR5, WG1, 11.1). Simulations run in a climate projection are "not intended to be forecasts of the observed evolution of the system but to be possible evolutions that are consistent with the external forcings." (IPCC, AR5, WG1, 11.1, p. 959)

What does this mean? Remember that a forcing is a perturbation that is external to a system, defined in some way that can potentially drive the system out of its range of natural variability and into a new one. The kinds of forcings that climate scientists are interested in are perturbations in the degree of energy from radiation that the climate system receives and retains due to changes in albedo, aerosols, solar variation, and, of course, changes in the molecular composition of the atmosphere that affect its emissivity (like concentrations of carbon dioxide). Ideally, climate scientists are interested in knowing how a climate system like ours would respond to various different external forcings (a) in terms of changes in statistical variables that summarize the natural variation and (b) that are *more or less independent of the particular state of the atmosphere, etc. that the system happens to start in*. It's a non-trivial question how easily this can be done – in fact, sorting out which features of a *climate simulation run* are consequences of the chosen initial condition is hard. The *hope*, however, is that climate scientists can do this in a way that avoids the problems that they face in predicting the weather beyond about a week, by focusing on questions that are independent of the features that weather models are keen to track: day to day internal variations of the system. We will come back in more detail to the question of how much these kinds of "experiments on climate forcings" depend on knowledge of initial conditions in Chapter 5.

What kinds of "climate experiments" are climate scientists interested in performing? One major area of interest is obviously in determining climate sensitivity: both the equilibrium value and transient responses on the order of decades. Equilibrium climate sensitivity is defined in terms of the very long-term average of the global surface temperature, and the transient response is usually something on the order of a multi-decadal average. Both measure a number that is averaged over large spatial and temporal scales. But we can see that climate projection does come in two varieties: the kind that is concerned with decadal averages and the kind that is concerned with genuinely time-scale-independent equilibrium values.

In addition to projecting climate sensitivity, climate simulations are also used in so-called fingerprinting studies. These are studies that aim to forecast the distinctive "fingerprints" that different kinds of forcings will imprint on the climate. They do this in order to attribute causes to measured changes in the climate over the last century. The IPCC report summarizes the idea like this: Attributing the cause of a change in the climate to one factor rather than another requires showing that the change is "consistent with results from a process-based model that includes the causal factor in question, and inconsistent with an alternate, otherwise identical, model that excludes this factor ... We cannot observe a world in which either anthropogenic or natural forcings are absent, so some kind of model is needed to set up and evaluate quantitative hypotheses: to provide estimates of how we would expect such a world to behave and to respond to anthropogenic and natural forcings." (IPCC, AR5, WG1, Executive Summary, pp. 872–873)

Models suggest, for example, that depletion of ozone in the stratosphere (one kind of forcing) makes the stratosphere colder, since ozone decreases emissivity and can therefore trap heat wherever it is found. The models show that volcanic eruptions (another forcing), on the other hand, add radiation-absorbing particles to the stratosphere, making it warmer. But the effects of volcanic eruptions aren't exactly the opposite of the effects of ozone depletion, because they also add reflective particles to the tropo-sphere, making it colder. Models show that an increase in solar forcing would warm the atmosphere from top to bottom. Greenhouse gas forcings, on the other hand, show a warming of the troposphere and cooling of that stratosphere that is very consistent with what has actually been observed in the post-industrial period. The fact that only the last "forcing experiment" agrees with observations of the real world is a substantial part of the evi-dential base for the claim of anthropogenesis.

Those are the two principal ways in which climate models are used to answer the "big questions" about climate (climate sensitivity and attribu-tion of observed trends), but we are also interested in other more fine-grained climate projections as well. What would various forcings do to such climate variables as sea-level, ice-melt, patterns of precipitation, ocean currents, tropical storms, etc.? In 50 years? In 100 years? Where will it flood? Where will there be drought? Will the average warming be evenly distributed, or more pronounced at the poles? Etc. Of course, it is well understood and acknowledged that as the projections made by climate models get more and more fine-grained, either temporally or spatially, their reliability goes down – in some cases, like the patterns of ice-melt, rather dramatically.

## 4.4 Features of Computer Simulations of the Climate

### *a) Equations*

We remarked above that computer simulation often involves taking models that comprise a set of partial differential equations and then discretizing those onto a grid. This is precisely how the core of climate models is constructed. Part of the core of a modern "flagship" climate model is a general circulation model of the atmosphere ("AGCM"). (See Boxes 4.1 and 4.2 for other variants.) The atmospheric model is a discret-ization of the following equations, which are a combination of a version of the Navier-Stokes equations and an equation for thermal effects of solar radiation[3]:

---

[3] In the equations, $r$ is the distance from the earth's center, $\Omega$ is the angular velocity of the earth's rotation, $\varphi$ is latitude, $\lambda$ is longitude and $t$ is time. $c_{pd}$ is the specific heat capacity of air at con-stant pressure, $\theta$ is potential virtual temperature, $\Pi$ is the "Exner function" of pressure and $\rho$ is air density. The subscript "d" refers to dry air. The directions are east-west ($u$), north-south ($v$) and vertical ($w$).

$$\frac{D_r u}{Dt} - \frac{uv \tan \phi}{r} - 2\Omega \sin \phi v + \frac{c_{pd}\theta}{r \cos \phi} \frac{\partial \Pi}{\partial \lambda} = -\left( \frac{uw}{r} + 2\Omega \cos \phi w \right) + S^u$$

$$\frac{D_r v}{Dt} - \frac{u^2 \tan \phi}{r} + 2\Omega \sin \phi u + \frac{c_{pd}\theta}{r} \frac{\partial \Pi}{\partial \phi} = -\left( \frac{vw}{r} \right) + S^v$$

$$\frac{D_r w}{Dt} + c_{pd}\theta \frac{\partial \Pi}{\partial r} + \frac{\partial \Pi}{\partial r} = -\left( \frac{u^2 + v^2}{r} \right) 2\Omega \cos \phi u + S^w$$

$$\frac{D_r}{Dt}(\rho_d r^2 \cos \phi) + \rho_d r^2 \cos \phi \left[ \frac{\partial}{\partial \lambda}\left( \frac{u}{r \cos \phi} \right) + \frac{\partial}{\partial \phi}\left( \frac{v}{r} \right) + \frac{\partial w}{\partial r} \right] = 0$$

$$\frac{D_r \theta}{Dt} = S^\theta$$

These equations may look daunting, but they are actually rather simple. The first three equations are just Newton's second law used to predict the acceleration of the winds in the east-west ($u$), north-south ($v$) and vertical directions ($w$). The fourth equation, in effect, guarantees that mass is conserved even though the density, speed and direction of the atmosphere changes as it flows around the earth. And the last equation, the thermodynamic equation, parameterizes heat-transfer processes, like the heating by the sun, with the term $S$. We use fundamentally similar equations to model the dynamics of the ocean, but usually make further simplifying approximations.

## b) Discretization

The typical size of the "grid" on which we discretize these equations is presently on the order of cubes of 100km on each side (though models exist that run at much higher resolution – and methods exist for making spatially higher resolution predictions, see Box 4.2), and is broken up into about 20 vertical layers. The exact form of the grid varies. Sometimes it is a polar grid, and sometimes computational efficiency is improved with other, more complex, schemes.

In sum, the model of the atmosphere consists of a three-dimensional grid of cells, with each cell exchanging radiation, heat, moisture, momentum, and mass with its neighbors.[4]

[4] (Stone and Knutti 2010)

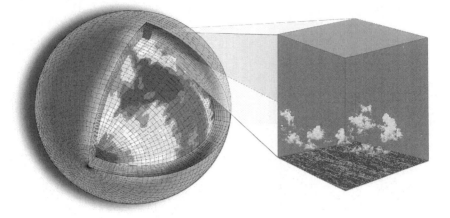

Figure 4.1 A typical discretization grid for a global climate model.
*Source*: adapted from www.gfdl.noaa.gov/pix/model_development/
climate_modeling/climatemodel.png

---

**Box 4.1  Earth Models of Intermediate Complexity (EMIC)**

EMICs are usually said to "bridge the gap" between energy balance models and full-blown high-resolution AOGCMs or ESMs. They are typically much lower resolution than GCMs and sometimes parameterize features like the climate sensitivity, or the degree of carbon feedbacks, that are resolved by more complex models. Thus they combine elements of GCM and energy balance models – indeed a variety of them exist that span the spectrum between those two extremes. They are used for much longer (e.g. thousands of years) simulations, especially for paleoclimate simulations, or for exploring different model components, couplings or ranges of parameters.

---

### c) Coupling

An AGCM can use off-the-shelf ocean temperatures as boundary conditions, or, as is the case in a modern flagship model, it can be coupled to a circulation model for the oceans. (This makes it a "coupled atmosphere-ocean general circulation model" or "AOGCM".) The model of the ocean is similar, but uses a simplified form of the above equations. It consists of a three-dimensional grid of cells, each cell exchanging radiation (at the top levels), heat, salt, momentum, and mass with its neighbors. Grid cells of the oceans tend to be about half the linear size of atmospheric

---

**Box 4.2 Downscaling Climate Models**

The typical horizontal grid size for an AOGCM or ESM at the present time is around 100km (see Box 4.3). "Downscaling" refers to the process of trying to extract more spatially fine-grained climate predictions out of such models. There are many different ways of achieving this goal, but they can be roughly separated into the dynamical and the non-dynamical. The latter involve, in one way or another, estimating higher resolution data from the output of the global models using mathematical methods of varying degrees of complexity: by extrapolation, statistical methods, or empirical methods. The former involve using physically based dynamical models that resolve at higher resolutions than ordinary GCMs, often by nesting them inside the grid cells of the GCMs. The GCM provides boundary conditions, on time intervals of several hours, for a regional or "limited-area" model that runs at a much higher resolution.[5]

---

grid cells. The vertical depth of the cells in each model tends to vary, with the vertical resolution getting smallest where the ocean interfaces with the atmosphere (or at any interesting interface). The time step for a typical climate model is about ten minutes, although the exchange of some quantities (for example radiation) between grid cells happens less frequently to save computer power. If an AOGCM is further coupled to a land model or a land ice model or a carbon cycle model, it is usually called an "Earth System Model" or "ESM." On the best available computers, it takes about one day of computer time to calculate one year of "earth time" for these models. This means a one-century run of an ESM can take many months to complete.

### d) Parameterization

If the most important feature of a climate simulation model is that it discretizes a set of continuous equations onto a grid, the second most important feature is how it treats interactions in the system that occur below the scale of the chosen grid. Anything that happens "within" a grid box cannot be calculated using those fundamental equations that have been discretized at that grid size. Instead, such an interaction needs to be treated with a "sub-grid model," or what is also called a "sub-grid parameterization" or just a "parameterization."

---

[5] For a good discussion of downscaling, and how much predictive skill these methods add to ordinary climate modeling, see Mearns et al. (2014)

Sub-grid modeling is not new to climate science. It goes back to the earliest attempts by John von Neumann and his colleagues to use computers to simulate fluid flows in bomb-design problems. In the study of turbulence in fluids, for example, a common practical strategy among engineers is to account for the missing small-scale *vortices* (or *eddies*) that fall inside the grid cells. This is done by adding to the large-scale motion an *eddy viscosity* that characterizes the transport and dissipation of energy in the smaller-scale flow – or any such feature that occurs at too small a scale to be captured by the grid. So "parameterization" in climate science refers to the method of replacing missing processes – ones that are too small-scale or complex to be physically represented in the discretized model – by a more simple mathematical description. This is as opposed to other processes – e.g. large-scale flow of the atmosphere – that are calculated at the grid level in accordance with the basic theory; these are said to be "resolved" as opposed to "parameterized." It is called "parameterization" because various non-physical *parameters* usually need to be filled in to drive the highly approximative algorithms that compute the sub-grid values. The parameters are "non-physical" (see Box 4.3) because no such value exists in

---

### Box 4.3  A Tale of Two Parameters

Consider two parameters you might use in building a model of how skydivers behave when they jump out of planes on this planet: "g," the acceleration of gravity, and "terminal velocity," the velocity at which a person falling freely near the earth's surface will stop accelerating. Neither one is a fundamental constant. The first one is a consequence of the laws of gravitation, but also of the mass and radius of the earth, etc. It is equal to roughly $9.8 \text{ m/s}^2$. The second one is a consequence of all of those, plus some facts about the atmosphere, the shape of human bodies, etc. If you look up the value of the second parameter, you will find that it depends crucially on a number of factors. A skydiver who holds her limbs out and presents her belly to the ground has a speed of around 125mph. A more experienced skydiver who tries to stand straight and point her feet to the ground will generally reach speeds of around 180mph. A highly practiced specialist at freefalling can reach speeds of 300mph. So what would be the "correct" value of terminal velocity to use in a simulation that involved skydivers? There really might not be a matter of fact about this. In the absence of stable facts about what percentage of the skydivers in my simulation were going to adopt each of the three positions, and some way of relating that to other variables in my simulation, "terminal velocity" might just end up being a kind of placeholder parameter. In such a case, it would make more sense to talk about the "best" value of such a parameter (in the sense of relativized adequacy for purpose), then it would to talk about the "correct one."

nature – both the need for the parameterization and the best value for the associated parameters, are artifacts of the computation scheme.

Two of the most important parameterizations in climate models involve the formation of clouds and the vertical exchange of water in the oceans, respectively. In an atmospheric model, all condensation processes, both the formation of rain and snow, and the formation of clouds, are represented by such parameters. But cloud formation is extremely important because clouds are important drivers of both the reflection of sunlight and the absorption of emitted infrared radiation. So changes in clouds in response to changes in climate can be an extremely important feedback. In the ocean model, eddies (or "vortices") that are smaller than the grid size, are responsible for moving much of the heat and mass around, and the rate at which the oceans can absorb heat is also extremely important since in fact most of the extra heat that greenhouse gases trap ends up in the oceans. To track the transport of heat and mass by sub-grid-sized eddies, climate models resort to some of the same methods as engineers, just on a larger scale.

Here is an example of a relatively simple parameterization that we are fairly confident works pretty well: clear sky radiative transfer – i.e. what happens to radiation as it passes through cloud-free regions of the sky. The scheme used to calculate this in the GFDL (see Box 4.4) model works something like this: it divides the relevant region of the electromagnetic spectrum into 18 bands; the emission and transmission rate for each band through the grid cell is then calculated from an equation based on the concentrations of several aerosols and gas components, including: sea salt, dust, black and organic carbon, and sulfate (aerosols) and $H_2O$, $CO_2$, $O_3$, $N_2O$, $CH_4$, CFC-11, CFC-12, CFC-113, HCFC-22 (gases).

Cloud parameterizations, on the other hand, are much more complicated and inspire somewhat less confidence. Once upon a time, cloud parameterization schemes simply calculated the percentage of a grid cell filled with clouds as a mathematical function of a few grid variables like humidity and convection. But clouds form via complicated interactions between large-scale circulations, convection, small-scale turbulent mixing, radiation, and microphysical processes like precipitation (including the impact on this of aerosols), melting and freezing, and nucleation of ice crystals. Different kinds of cloud systems are driven by very different kinds of considerations. So modern cloud schemes involve parameterizing some or all of these different effects and jointly calculating their contribution to cloud physics.[6] It is a very complex endeavor and it is widely agreed that a

---

[6] (Chaboureau and Bechtold 2002)

---

**Box 4.4  GFDL CM2.X**

Geophysical Fluid Dynamics Laboratory Coupled Model, version 2.X is an AOGCM. It was developed at the NOAA Geophysical Fluid Dynamics Laboratory in the United States and was one of the main climate models contributing to CMIP3 and the fourth Assessment Report of the IPCC. It also contributed to CMIP5 and the AR5 (though its importance was partly supplanted by CM3). The grid size of the atmospheric component is 24-levels in the vertical, and has horizontal resolution of 2 degrees in the east-west (111km at the equator) and 2.5 degrees in the north-south directions. The grid size of the oceanic component is 50 levels in the vertical and 1 degree in the east-west direction. It varies from 1 degree in the polar regions to 1/3 of a degree along the equator in the north-south direction.

The computational model itself contains over a million lines of code. There are over a thousand different parameter options. It is said to involve modules that are "constantly changing," and involve hundreds of initialization files that contain "incomplete documentation." The CM2.X is said to contain novel component modules written by over one hundred different people. Just loading the input data into a simulation run takes over two hours. Using over a hundred processors running in parallel, it takes weeks to produce one model run out to the year 2100, and months to reproduce thousands of years of paleoclimate. If you store the data from it every five minutes, it can produce tens of terabytes per model year.[7]

---

great deal of the variation in the outputs of different climate simulations is attributable to differences in the ways in which they treat clouds.

### e) Modularity

The software architecture of most GCMs/ESMs is highly modular. "Modularity," roughly, is the degree to which a system's sub-components can be separated and recombined. It is a key insight, going back at least to Herbert Simon, that complex software design and quality control are greatly aided by modularity because, at least in principle, it allows the implementation, maintenance, and assurance of reliability of a complex system to be broken down and handled vis-à-vis smaller units. Again in principle, model modularity allows the possibility of establishing the credibility and reliability of each of the modules separately.

---

[7] All of the above claims and quotations come from Dunne (2006)

In order to study the degree of modularity in climate models, computer scientists Kaitlin Alexander and Steve Easterbrook conducted a study in which they analyzed the source code of many of the models that participated in CMIP5. They then created some very beautiful and elegant software diagrams that show how the models are broken up into components, from a software architecture point of view, how they are interconnected, and how large each module (in terms of lines of code) is relative to each other, and how they are coupled (Alexander and Easterbrook 2015).

Interestingly, they found that the major climate models can be divided into two different types in terms of their software architecture. They call them "star-shaped" and "two-sided" models, respectively. The former are more common in the United States and the latter are more common in Europe. Star-shaped models are more modular. They have four or five different modules (out of: atmosphere, ocean, land, sea ice, land ice), each entirely independent and communicating only with a central coupler. The two-sided models only have a central coupler between the main module of the ocean and main module of the atmosphere, and the other modules are subsidiary to either the ocean or the atmosphere.

Ideally, as Alexander and Easterbrook note, we would like our climate model modularities to "carve nature at the joints." (Alexander and Easterbrook 2015). We would also, of course, like them to carve our scientific teams into communities of expertise. In practice, they probably succeed better at the latter than the former, because the massive complexity of the climate doesn't allow for as much nature-carving as we might otherwise like. It is important to stress, therefore, that the architecture that is displayed in these pictures is the software architecture of the computer programs, and not, in any deep sense, either the natural architecture of the climate system itself or the scientific architecture of the models.

There are a few different reasons why overemphasizing the software-architectural modularity of climate models can be somewhat misleading, particularly if we think of modularity in terms of both (a) the degree to which a system's sub-components can be separated and recombined and (b) the affordance of being able to break down the implementation, maintenance and assurance of reliability of the model into smaller units.

The first reason is the most fundamental, and it has to do with the highly complex non-linear interactions between what we take to be the different components of the climate system, particularly between the ocean and the atmosphere. James Randall makes the point quite well here:

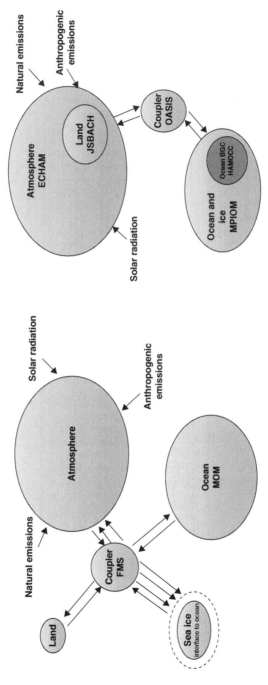

**GFDL** climate model 2.1 (coupled to MOM 4.1)
Geophysical fluid dynamics laboratory, USA

Solar radiation

Anthropogenic emissions

Atmosphere

Natural emissions

Ocean
MOM

Coupler
FMS

Land

Sea ice
(interface to ocean)

**COSMOS** 1.2.1
Max-Planck-institut für meteorologie, Germany

Natural emissions

Anthropogenic emissions

Atmosphere
ECHAM

Land
JSBACH

Coupler
OASIS

Ocean and ice
MPIOM

Ocean BGC
HAMOCC

Solar radiation

Figure 4.2 A star-shaped, and a two-sided climate model, respectively.
*Source:* adapted from Alexander and Easterbrook 2015

> Climate models have many coupled components. From a management as well as software engineering viewpoint, the concept of plug-compatible model components, or modules, is almost irresistibly attractive. From a physical viewpoint, however, the possibility of modularity is illusory because the physical climate system itself isn't modular. In fact, coupling model components is one of the most important, subtle, and neglected aspects of model development. Here's a timely example: cumulus clouds are strongly coupled to the turbulent boundary layer near the ground, and this coupling is crucial for the day-night cycle of precipitation over land. In many models, however, Scientist A formulates the cumulus component, Scientist B develops the boundary layer component, and the coupling between them falls through the cracks – which might explain why models have trouble simulating the day-night cycle of precipitation over land. The point is that the illusion of modularity discourages modelers from giving an appropriate amount of attention to the physical couplings among climate processes. (Held and Randall 2011)

The climate system permits a certain degree of physical independence between its components. But sometimes, the degree of software modularity in climate models exceeds the degree of physical independence that the actual system permits. Johannes Lenhard and I argued that the modularity of climate models ought to be considered "fuzzy," in a way that makes them epistemologically opaque (Lenhard and Winsberg 2010). We will return to these issues in both Chapters 9 and 10.

A second basic reason has to do with parameterizations. We would like to think of parameterizations as being associated with a correct parameter value that gives them their own *independent degree of skill*. Unfortunately, they are not. They should be properly understood as components that add to or diminish the *overall skill* of a particular modeling package with which they are used in conjunction. What might independently look like an improvement in a cloud parameterization scheme will not add to overall model skill if the model's planetary boundary layer parameterization was designed and tested alongside an older cloud scheme. Many different groups of parameterizations work synergistically in this sort of way.

### f) Tuning

Finally, climate models require tuning (Mauritsen et al. 2012). The term "model tuning" is not always well defined, but I am using it here to mean the process of adjusting a few key parameter values (associated with the sub-grid parameterizations discussed above) in order to get the behavior of the model *as a whole* to behave in an acceptable way. This happens as a last step, after model structure has been chosen, and after sub-grid parameterizations are chosen and refined for reasons intrinsic to their individual performance. The first thing that a climate simulation absolutely has to get right is the top-of-the-atmosphere (TOA) energy balance. Remember the energy balance models of Chapter 3. They predict global warming/cooling whenever there is an imbalance between the amount of solar radiation coming into the system at the TOA and the amount of emitted radiation leaving. To have much credibility about anything related to this sort of imbalance, therefore, any climate simulation had better get values for TOA energy imbalance that match observations. The first step in climate model tuning, once the model is put together as a whole, is thus to adjust key parameter values until the TOA energy balance is deemed to be correct. Model tuning also can be used to address other objectives, including getting a match with observed global mean temperature, with the general features of atmospheric circulation, tropical variability and sea-ice seasonality. Tuning is a feature of climate models that is deep and fundamental to understanding their evaluation and sanctioning, so we will return to it in more detail later in that context.

## Suggestions for Further Reading

Winsberg, Eric. 2010. *Science in the Age of Computer Simulation* (especially Chapters 1–3). A monograph on philosophical issues related to computer simulations, mostly in the physical sciences.

2013. "Computer Simulations in Science," *The Stanford Encyclopedia of Philosophy*. A broader survey of philosophical work on computer simulation in the sciences, including the biological and social sciences. Contains an extensive bibliography.

Mauritsen, Thorsten et al. 2012. "Tuning the Climate of a Global Model," *Journal of Advances in Modeling Earth Systems*. An extremely useful discussion of a topic that is not very often discussed in the scientific literature; more or less the *locus classicus* on climate model tuning.

Lenhard, Johannes and Eric Winsberg. 2010. "Holism, Entrenchment, and the Future of Climate Model Pluralism," *Studies in History and Philosophy of Modern Physics*. A paper on some of the epistemological implications of the "fuzzy modularity" of climate models.

Held, Issac and David Randall. 2011. "Point/Counterpoint," *IEEE Software*. A debate about the possibility of "open source" climate modeling software. Especially interesting in the context of this chapter for its discussion of the difference between scientific modularity and software modularity.

Alexander, Kaitlin and Steven Easterbrook. 2015. "The Software Architecture of Climate Models: A Graphical Comparison of CMIP5 and EMICAR5 Configurations," *Geoscientific Model Development*. A very nice analysis of the software modularity of climate models with some beautiful diagrams illustrating their various forms.

# 5

# *Chaos*

## 5.1 Introduction

Everyone knows that the weather is difficult to predict for long periods of time. Most of us know that this has something to do with the fact that the dynamics of the atmosphere are complex, non-linear, and almost certainly *chaotic*. These three features of the atmosphere, especially the fact that it is chaotic, have obvious epistemological consequences for weather prediction. They also have consequences for climate science, but the differences between climate and weather make these latter consequences, especially the consequences of chaos, somewhat more complicated. The chaotic behavior of the atmosphere makes long-term weather prediction nearly impossible. As we will see, this is because a small degree of uncertainty concerning the precise present conditions of the atmosphere can quickly turn into a large degree of uncertainty about its future conditions. This, in a nutshell, is what is popularly known as "the butterfly effect." Chaos gives rise to epistemological problems for climate science as well, and climate scientists sometimes talk about one of these problems using the expression "initial condition uncertainty." This can easily make one think that chaos affects climate science in the same way in which it affects weather prediction. But this is much too simple, and it can easily lead to some confusions about what the epistemological consequences of non-linearity, complexity, and chaos are for climate science. Some of these confusions have made their way into the philosophical literature. We should start, then, with a brief introduction to chaos and some of its characteristic features.

## 5.2 What Is Chaos?

The climate system as a whole is almost certainly chaotic, and so any student of climate science is well served by having at least a rudimentary understanding of chaos. So, what is a chaotic system, exactly?

---

**Box 5.1\*  A Formal Definition of a Dynamical System**[1]

Some of the philosophical literature employs slightly more formal description of a dynamical system. This description treats the state space as a manifold[2], and the evolution function as either a *map* from the manifold back onto itself or as a function from the manifold and a time parameter back to the manifold (called a "*flow*"). We use the first formalism in the case of an evolution function that is discrete in time, and the latter in the case of an equation that is continuous in time. In the first case, we represent it as:

$$\phi : X \to X$$

This says "$\phi$ is a function from X onto itself. In the second case, we represent it as:

$$\phi : M \times \mathbb{R} \to M$$

This says "$\phi$ is a function from the manifold M and the real number line (to represent time) back onto the manifold." In other words, if you give me a point in the manifold, and an element, t, from the real number line representing an amount of time, $\phi$ will tell you where, on the manifold, you end up after time t.

---

Let's start with a fairly abstract mathematical characterization of what it is to be a *system*. Dynamical systems theory begins with the idea of a "dynamical system," an abstract entity used for representing systems of interest. A dynamical system is an abstract entity whose state at any moment in time can be completely characterized by the value of a set of "state variables" $\{x_1, x_2, x_3 \ldots x_n\}$. Dynamical systems thus "live" in a space of states that represents all the different possible ways the system could be at any instant in time. The system's dynamics – the manner in which the system moves from state to state over time – is specified by a set of equations of evolution that either a) give the rate of change of each variable as a function of the state variables (that is, via a set of differential equations) or b) specify the value of each of the state variables at successive discrete moments in time as a function of their values at the previous moment (that is, via a set of difference equations). The equations of evolutions thus

---

[1] Boxes marked with an "\*" after the box number are for mathematically advanced readers. They can be skipped without affecting comprehension of the rest of the book.

[2] A manifold is a topological space that is locally Euclidean. Simple examples of manifolds are objects like the real line, a plane surface, or the surface of a sphere. (The first two are Euclidean–period, but the last one is only locally, but not globally Euclidean.)

pick out allowable trajectories through the space of states (or "state space"). We are especially interested in systems that are deterministic. Being deterministic means that the equations of evolution in question specify a *unique trajectory* through the space of state variables for some given set of initial conditions. Chaotic systems are first of all, by definition, deterministic dynamical systems.

We can ask the following questions about any deterministic dynamical system: if we run the equations of evolution on two nearby points, do they stay nearby as we move ahead in time? Or do they move apart? If they move apart, how quickly do they diverge? There are a variety of possible answers. Suppose for example our system is a simple harmonic oscillator (let's say a mass, m, attached to a spring obeying Hooke's law: with a linear restoring force equal to $-kx$), described by the differential equation:

$$m\frac{d^2x}{dt^2} = -kx \tag{5.1}$$

When we have differential equations specifying a system's evolution through state space, it is not always the case that we can write down an explicit solution to the equations, e.g. some simple polynomial or trigonometric[3] function of time that spits out the values of each state variable for any given time. But in the case of equation 5.1 we can. The set of solutions is:

$$x_t = \varphi(t, x_0) = A\cos(\omega t + x_0) \tag{5.2}$$

where $\omega = \sqrt{\dfrac{k}{m}}$, and $A$ and $x$ depend on the initial conditions (the starting position and velocity).

---

**Box 5.2\* Formal Notation Continued**

It should be clear that equation 5.2 has the form

$$\phi : M \times \mathbb{R} \to M$$

Where in our case the manifold M is the real number line representing the possible positions of the mass.

---

[3] See en.wikipedia.org/wiki/Closed-form_expression for a useful discussion of the various ways this sort of condition can be made more precise.

If we had two identical such systems, and we started them at the same time with slightly different initial conditions (either a different initial position or initial velocity, or both) the amount by which the two systems differed would not change very much over time.

If we add a damping force to the system, let's say because the mass on the end of the spring has to slide along a frictional surface, we can have an equation of evolution with an added term:

$$m\frac{d^2x}{dt^2} = -kx - c\frac{dx}{dt} \tag{5.3}$$

(assuming the force of friction is proportional to the velocity of the mass).

As long as the so-called damping ratio $\left(\dfrac{c}{2\sqrt{mk}}\right)$ is less than one, the system will behave more or less in the same way as its undamped cousin, except the amplitude A decays exponentially with a decay rate of c/2.

The set of solutions of the differential equation is:

$$\varphi(x_0, t) = ae^{\frac{-vt}{2}} \cos(\omega_1 t - x_0) \tag{5.4}$$

The mass will oscillate back and forth, but lose amplitude with every oscillation, and will eventually converge on the point x=0, regardless of what initial condition we start with. We can ask the same question as above: if we had two identical such systems, and we started them at the same time with slightly different initial conditions (either a different initial position or initial velocity, or both) what is the amount by which the two systems will come to differ over time? This time the answer is that the amount will decrease over time, quickly approaching zero, as every system converges on a point (the "attractor" – see below) where the velocity is zero and the position is the point where the spring is neither compressed nor extended.

In sum, one very general way in which we can characterize a dynamical system is by describing the rate at which two nearby systems diverge over time. One of the hallmarks of *chaotic* systems is that nearby systems move away from each other *very quickly* (at least for some period of time). The expression used to describe this state of affairs is "sensitive dependence on initial conditions," (SDIC) or sometimes, the "butterfly effect." (See Box 5.1 for an explanation of the difference between a chaotic system and one that (merely) exhibits SDIC.) There are different ways of defining SDIC, but one that works well for our purposes is to say that a system has

SDIC just in case the rate of growth in errors is exponential in time (this is sometimes called "exponential SDIC, but since it's the main kind that interests us, we will sometimes just call it "SDIC"). That is, the amount we expect two nearby systems to have diverged after some amount of time t is proportional to the power of t. More carefully, we can say that a system displays SDIC in this sense if the rate at which nearby systems move away from each other (at least to begin with), in the space of possible states they can be in, is exponential in time. That is, if $\delta Z(0)$ is the initial distance that separates two such systems in the state space and $\delta Z(t)$ is their distance at time t, then we say that the systems display SDIC just in case:

$$|\delta Z(t)| \approx e^{\lambda t} |\delta Z(0)| \qquad (5.5)$$

where $\lambda$ is the "Lyapunov exponent"[4] and e is Euler's number = 2.71828.... We can see that as $\lambda$ grows, the rate at which errors grow goes up exponentially.

The inverse of the Lyapunov exponent $(1/\lambda)$ is the Lyapunov time. Notice that the larger lambda is, the faster two nearby systems move apart. This makes lambda extremely important for the notion of initial condition uncertainty. It tells us something very important about what it will

---

**Box 5.3\* SDIC in Formal Notation**

Using the formalism we described above, we can define SDIC a bit more formally.

Let $\phi : X \rightarrow X$ be a time-evolution function on the manifold X. Say a dynamical system's behavior is *exponentially sensitive to initial conditions* if there exists $\lambda > 0$ such that for any state $x \in X$ and any *delta* > 0, almost all elements $y \in X$ satisfying $0 < d(x, y) < \delta$ are such that $d(\phi^t(x), \phi^t(y)) > e^{\lambda t} d(x, y)$.

That's if $\phi$ is a map. If $\phi$ is a flow, then the definition becomes:

Let $\phi : M \times \mathbb{R} \rightarrow M$ be a time-evolution function on the manifold X. Say a dynamical system's behavior is *exponentially sensitive to initial conditions* if there exists $\lambda > 0$ such that for any state $x \in X$ and any *delta* > 0, almost all elements $y \in X$ satisfying $0 < d(x, y) < \delta$ are such that $d(\phi(x,t), \phi(y,t)) > e^{\lambda t} d(x, y)$.

---

[4] In fact, a single Lyapunov exponent is a fairly large simplification for a multi-dimensional system. For most systems, the value for the exponent depends on both on the starting point and on the orientation (in however many dimensions the state space has) of the initial separation vector. So strictly speaking, a system really has a whole spectrum of Lyapunov exponents for every point in its state space. Sometimes the "maximal Lyapunov exponent" is used as the best indicator of the predictability of the system as a whole – since it gives a kind of "worst case scenario" for predictability.

mean to our predictions of the future of the system if we have some small degree of error in our estimate of the initial condition of the system. The Lyapunov time, the inverse of the exponent, gives us a ballpark figure for the timescale on which we can expect to be able to make useful predictions of a certain kind for a system that displays this kind of SDIC. It is widely accepted that our best weather prediction models will have Lyapunov times of around three days. What this means, of course, is that if you had the exact state space of the earth's weather, and you knew its initial condition in that space to within some range $\delta$, the amount of error you would expect to have in your prediction of where the system would be in three days would be multiplied by e (about 2.7 in decimal notation), and after six days by e^2, and after 12 days by e^4 (see equation 5.5). This means that your degree of uncertainty about the system after 12 days would have risen to about 55 times $\delta$. This gives a back-of-the-envelope explanation of the fact that weather predictions tend to be poor after about five-six days and useless after ten.

### 5.3 Two Toy Chaotic Models

It will be useful at this point to take a closer look at two famous chaotic systems: the logistic map and the Lorenz model. The second of these, like our examples above, has a set of differential equations that describe its evolution, and therefore specifies a flow. The first has a single difference equation that is discrete in time, and therefore picks out a map.[5]

A) The logistic (family of) map(s) is a set of very simple, one-dimensional, quadratic mappings, given by the parameterized equation

$$x_{n+1} = rx_n(1 - x_n) \tag{5.6}$$

with the parameter $r$ taking values on the interval (0,4] and the initial value for $x_0$ being set to a number between 0 and 1.

The logistic map exhibits a variety of different behaviors, depending on what value we give to $r$[6], but for most of the values between about 3.57 and 4, the behavior of the map has all of the hallmarks of chaos, including exponential growth, with respect to time, in the distance between two

---

[5] Interestingly, the Poincaré-Bendixson theorem tells us that a dynamical system governed by continuous equations can be chaotic only if it lives in three or more dimensions, whereas one governed by a difference equation can be chaotic at any dimensionality at all.

[6] The behavior of the family of logistic maps as r moves from 0 to 4, in which it undergoes what is known as the "period doubling cascade," and which in some sense explains the onset of chaos, is itself fascinating.

---

**Box 5.4 Chaos vs. (Mere) SDIC**

SDIC, even in the form of exponential error inflation as defined in equation 5.5, is not a sufficient condition for chaos. Consider, for example, a system whose state at time t is given by the closed-form equation $x(t) = x(0)e^{\wedge}t$. We could set two initial values for $x(0)$ that were very close but the distance between them would grow with $e^{\wedge}t$ and hence would qualify as SDIC. But few people would want to count such a system as chaotic. Intuitively, the added feature that a system has to have to be chaotic is for its state to be tightly confined to a relatively small region of the state space while, *at the same time*, having exponential error growth. What makes this counter-intuitive pair of features possible at the same time is the existence of strange attractors that the system can live on, of which the Lorenz attractor is typical. How, exactly, chaos ought to be defined is a matter of some controversy, but one popular definition, due to Robert Devaney, combines the requirement of SDIC (a metrical property) with two topological properties: topological mixing and dense periodic orbits. (In most cases, the two topological properties imply SDIC – but since SDIC is, of the three criteria, the one of greatest practical importance, we don't leave it out of the definition even though it doesn't strictly speaking add anything.) Again intuitively, topological mixing says that if you start with any open set in the state space, and evolve forward the point in the set, at least some of those points will eventually take you into any other region you like. One way of thinking of this requirement is that it says that if you wait long enough, it will be irrelevant where you started for predicting whether or not you end up somewhere else. Having a dense periodic orbit means that every point in the space is approached arbitrarily closely by periodic orbits that are followed by the system.

Some systems are chaotic everywhere. But for many systems, especially those described by continuous differential equations, chaotic behavior is only found in a subset of the state space. Often, though, that set of points is an attractor for the system, meaning that a larger set of points in the state space is lead, via the dynamics, to the chaotic region eventually anyway. Thus, in such a system, chaos is in the future for a large set of points anyway.

---

initially close points. The Lyapunov exponent for the logistic map when it is chaotic is in the neighborhood of 0.5, depending on the precise value of r. At the same time, however, the behavior of the map is tightly confined to the region between 0 and 1 – which enables it to display both topological mixing.[7] (See Box 5.1.)

---

[7] The tight confinement means, of course, that the exponential growth in the distance between two initially close points can only go on for so long until they run out of room to diverge. After that, the distance between the points will start to more or less fluctuate at random. The best analogy for

---

**Box 5.5\*   Topological Mixing**

In the language of the formalism that we have introduced, topological mixing can be defined as follows:

A time-evolution function $\phi$ is called *topologically mixing* if for any pair of non-empty open sets $U$ and $V$, there exists a time $T > 0$ such that for all $t > T, \phi^t(U) \cap V \neq \emptyset$.

This, says, intuitively, that if I apply phi to all of the points in U for times longer than T, some of those points will end up in V, no matter what V is (that's what it implies to say that the intersection of the two sets is not the empty set). In other words, no matter where you are in the state space, if you wait T long, some of U will come to you. That's why it is called "mixing."

---

B) The Lorenz model is a three-dimensional, parameterized model (there are three different state variables that are each a function of time) given by a set of ordinary differential equations.

$$\frac{dX}{dt} = -\sigma X + \sigma Y \tag{5.7}$$

$$\frac{dY}{dt} = -XZ + \rho X - Y \tag{5.8}$$

$$\frac{dZ}{dt} = -XY - \beta Z \tag{5.9}$$

Here X, Y and Z are the state variables, t is time, and $\sigma, \rho$ and $\beta$ are parameters for specifying the particular version of the model we are interested in. So in the language of our formalisms above, this is a "flow." Edward Lorenz himself studied the model with values $\sigma = 10, \rho = 8/3$ and $\beta = 28$. For these values, the model is chaotic and has Lyapunov exponents of 0.90563, 0, and –14.57219, in the X, Y, and Z directions, respectively. (Meaning that errors grow especially fast in the x direction.)

The Lorenz model has an attractor[8]. An *attractor* is a region of the space of state variables toward which a system tends to evolve, for a wide variety

---

thinking about this behavior is to a baker stretching and folding his dough. Imagine that a fly lands on a sheet of dough and is cut in half. As the baker stretches the dough out, the two halves of the fly get farther apart. But eventually, he runs out of room and folds his dough back onto itself, invariably bringing the two fly pieces closer together. But then the stretching and folding continues, etc., until the distance between the two fly pieces is no longer interestingly related to their initial distance at all.

[8] Hereafter when I refer to the "Lorenz model" I mean the particular model with the parameter values $\sigma = 10, \rho = 8/3$ and $\beta = 28$.

of initial conditions. System states that get close enough to the attractor stay close to the attractor even if slightly perturbed. Chaotic systems like the Lorenz model tend to have "strange attractors" which have *fractal structure* due to the stretching and folding of the dynamics that is typical of chaotic systems.

These two models (the Lorenz model and the logistic map) allow us to illustrate a number of features of chaotic systems that are relevant to some philosophical discussions of climate systems. The first one that is worth illustrating is that, even in a chaotic system like the Lorenz model, the degree of predictability depends a great deal on the starting point:

Even though the maximum Lyapunov exponent gives us an estimate of how fast errors can grow, the predictability of a chaotic system depends on a number of factors. For the Lorenz class of model, it depends on the values of the parameters (some values are non-chaotic, and among the chaotic ones, the Lyapunov exponents vary); it depends on what initial value we start with; it depends on the variable of interest (the X variable is the one with the large exponent in the Lorenz model); and it depends on the kind of prediction we want to make!

## 5.4 Mitigating Butterflies

What do I mean by the *kind* of prediction? Look again at Figure 5.1, and suppose that you are starting with a point in the initial circle of image b. If you are uncertain where, exactly, the point you are starting with is in the circle, then any prediction you make about the value of X will quickly become useless. But suppose you think of that same circle as characterizing a probability distribution over a range of possible values; let's say one that is uniform over the volume represented by the circle.[9] You could then confidently predict that at some later time, the probability of finding the system in any particular state was given by the probability distribution that was uniform over the volume that is represented by the lowest smiley-face-shaped region at the bottom of image b. It is possible, in other words, to move from making point predictions to making predictions about ensembles as a way of mitigating part of the uncertainty created by a chaotic model.

Let's call this strategy for dealing with the butterfly effect "probabilistic initial condition ensemble forecasting" (PICEF) because it employs an ensemble of initial conditions over which we place a probability

---

[9] One dimension is obviously missing from the diagram, so the circle actually represents a sphere.

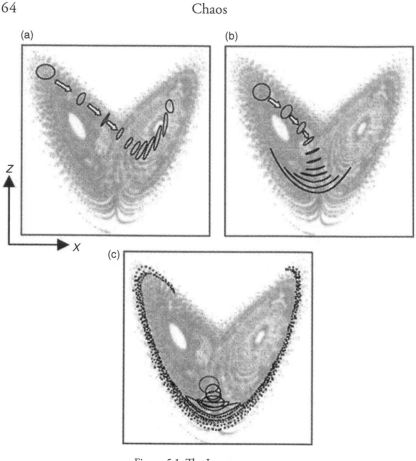

Figure 5.1 The Lorenz attractor.
Depending on the **starting point** on the attractor (the grey dust figure) there is
a) high predictability for a longer time
b) high predictability only in the near term, then increasing uncertainty
c) no predictability
*Source*: adapted from Slingo and Palmer 2011

distribution. We can see how PICEF is a useful way to mitigate some of the epistemic difficulties that the butterfly effect gives rise to. Even in the case where the first circle in image c represented our best knowledge of the system, we would not be reduced to having nothing useful to say about the system after the period of time depicted in the diagram. As long as we were confident in a probability distribution over that initial circle, we could confidently predict another probability distribution spread out over the necklace-shaped space of points.

What we have seen so far explains a great deal about why weather forecasts are probabilistic. Weather forecasters are often in a good position to narrow down the state of the atmosphere to some relatively small region of state space, and because the atmosphere is mixing (see Box 5.1), it is often reasonable to put a probability distribution over the small region. Probabilistic forecasting ensues.

Is this how climate scientists primarily deal with the butterfly effect? No. Climate science uses an entirely different strategy. Let's look again at the Lorenz model.[10] If we look back at Figure 5.1, we can see that there are at least two kinds of predictions we could make using the model that would keep us safe from the effects of SDIC, both of which are illustrated in the diagram. The first would be to restrict ourselves to the kinds of predictions illustrated in image a and to restrict ourselves to time scales over which the neighborhood we are interested in has not dissipated too much. The second would be to employ PICEF. Even if we are dealing with the situation depicted in image c, we can still put a probability distribution over the range of initial conditions we think are possible, and then allow the equations of evolution to forecast a future probability distribution – even if the latter is going to be quite scattered around the attractor.

But there is also a third kind of prediction we could make that is not at all illustrated in Figure 5.1. We turn instead to Figure 5.2.

Figure 5.2a shows a typical time series plot for the behavior of the X variable in the Lorenz model. That is, it shows the value of the variable X as a function of time from t=0 to t=16,000. We can also study another model that is closely related to the Lorenz model – one where we add a parameterized forcing to the model, by adding a factor "f" in the first two equations.

$$\frac{dX}{dt} = -\sigma X + \sigma Y + \boldsymbol{f} \tag{5.10}$$

$$\frac{dY}{dt} = -XZ + \rho X - Y + \boldsymbol{f} \tag{5.11}$$

$$\frac{dZ}{dt} = -XY - \beta Z \tag{5.12}$$

If we study this model – the Lorenz model *with an added forcing* (the part of the model written in bolder type, above) – we get the plot in

---

[10] Here I closely follow the presentation of Slingo and Palmer (2011)

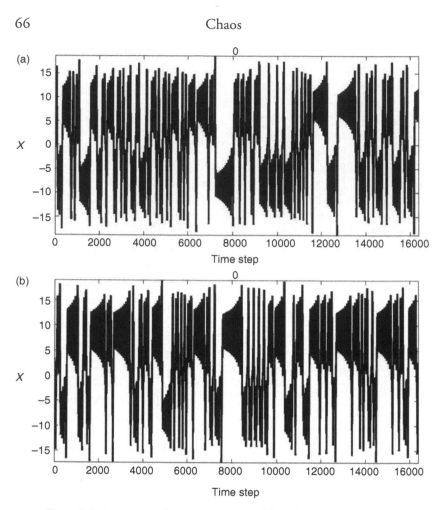

Figure 5.2 A time series diagram of the x variable in the Lorenz model with
a) no forcing and b) external forcing.
*Source*: adapted from Slingo and Palmer 2011

Figure 5.2b. Notice that the behavior of time series a and time series b are almost exactly the same. The only thing that has changed is the frequency of the upper pattern and the lower pattern. By experimenting with this class of models on a computer, you can show that this is a "predictable" effect of varying the forcing factor f. Varying the forcing factor doesn't change at all the number of the spatial patterns of regimes, but it does change how frequently they occur. There is more likely to be a positive X value in the second model. There is also more clustering in the second model. This is a very reliable prediction one can make about this family of Lorenz models. And notice that this kind of prediction is completely immune to the butterfly effect. PICEF can mitigate the effects of the butterfly effect, but forcing experiments can (almost – more on this in a bit) completely obviate them.

This point becomes particularly important if you recall what we learned about the primary purposes of climate models in Chapter 3. As we put the point there: climate models are not generally in the business of predicting climate variables *conditional on present conditions*. To the extent that the present weather conditions contribute to the future evolution of the climate system, they contribute to aspects of its *internal variability*, which is precisely what we are trying our best to leave out when we talk about climate and when we make climate *projections*. Recall the distinction that the IPCC makes between climate "predictions," (which the IPCC is much less concerned with) and climate projections, which it defined as: "climate change experiments with models that do not depend on initial conditions but on the history and projection of climate forcings." (IPCC, AR5, WG1, 11.1, p. 958) Simulations run in a climate projection are "not intended to be forecasts of the observed evolution of the system but to be possible evolutions that are consistent with the external forcings." (IPCC, AR5, WG1, 11.1, p. 958).

The kinds of forcings that climate scientists are interested in, recall, are perturbations in the degree of energy from radiation that the climate system receives and retains due to changes in albedo, aerosols, solar variation, and, of course, changes in the molecular composition of the atmosphere that affect its emissivity (like concentrations of carbon dioxide). Ideally, climate scientists are interested in knowing how a climate system like ours would respond to various different external forcings a) in terms of changes in statistical variables that summarize the natural variation and b) that are *more or less independent of the particular state of the atmosphere, etc., that the system happens to start in.* Climate projection is about learning about the outcome of forcing experiments, and as such does not rely much on PICEF.

Does this mean that projections, which are forcing experiments that don't depend on PICEF are *always free of initial condition uncertainty*? This is a very important question and it requires a response involving two subtleties.

1) First, how much any projection involves initial condition uncertainty depends on how well one can explore the family of models. Let's call

the different patterns of time series that one can see using the unforced equations, but depending on the initial conditions one chooses, the "natural variability" of the models. Both the unforced model and the forced model will have a degree of natural variability. If we see a difference in the behavior of two models, and we have adequately explored the degree of natural variability of each of them, we can safely conclude the difference is due to the forcing. In such a situation, there can be no initial condition uncertainty.[11] But if, for whatever reason, maybe because model runs are too time-consuming or expensive, we don't have a firm grip on the degree of natural variability of each model, we can't be sure if the difference we see is entirely attributable to the degree of forcing, or is an artifact of the initial conditions we chose. This gives rise to what climate scientists call "initial condition uncertainty." This shows two things. First, even these kinds of predictions – predictions about the effect of altering a forcing on the long-term statistical behavior of a system – can be subject to initial condition uncertainty. But second, in such cases, *initial condition uncertainty is not uncertainty about initial conditions*. In general, when climate scientists talk about initial condition uncertainty, they are not talking about uncertainty about what the actual initial conditions are. Rather, it is uncertainty about whether to attribute changes to the forcing or to natural variation when one has not studied all the possible behaviors of the systems. When climate scientists talk about initial condition uncertainty in, for example, determining climate sensitivity using a model, this is the kind of initial condition uncertainty they are talking about. Though it is appropriate to call it initial condition uncertainty, because we are uncertain if the change is due to the forcing, or to a *choice* of initial condition, it is not uncertainty about what initial condition *obtains in the real world*. The climate sensitivity is a prediction about the initial-condition-independent result of adding a forcing to the system.

2) Second, we have to remember, once again, that "climate is what you expect, weather is what you get." Recently, especially in the United Kingdom, projects have emerged that aim to make climate forecasts at the decadal scale and at very local spatial scales. The "United Kingdom

---

[11] Here of course we need to appropriately distinguish those model inputs we call initial conditions from those we call parameters. Obviously a projection will depend on values of variables like $CO_2$ concentration and insolation, but since those are not dependent variables we don't think of the model inputs for these variables as initial conditions.

Climate Impacts Programme" (UKCIP) is a (in)famous case of this.[12] Smith, Daron, and Stainforth, commenting on the UKCIP, have argued that if we want to make decadal forecasts, we need to take account not only of changes in the climate but also of internal variability, since variability will surely play at least as much of a role as climate change is likely to do over the course of a decade.[13] Let's call this strategy for making forecasts the "joint strategy." In other words, they argue that if you want to know what the hottest day in London will be in the 2050s decade, you need to know two things: how much will the climate have changed by 2050, and what will be the effect of internal variability on that particular decade vis-à-vis the expected climate for that decade. Both of those effects will contribute to the hottest day of the decade. The first thing we should note, of course, is that the joint strategy is not a method for making projections (in the technical IPCC sense). It is a method for making predictions.

## 5.5 The Joint Weather/Climate Strategy

If groups like the UKCIP were to take such a suggestion seriously and adopt the joint strategy, then uncertainty about initial conditions, even very small ones, would be susceptible to the butterfly effect. Indeed, it is hard to see how they would fail to be *swamped* by the butterfly effect. When a forecast (to use a neutral word) depends on initial conditions, and extends in time orders of magnitude more than the Lyapunov time, such a forecast is misguided.

The more sober and sensible conclusion, if all of the above is right, would seem to be that decision-guiding forecasts ought to be limited to climate *projections*, and that the joint strategy ought to be avoided – even if that means abandoning the idea of making projections that are relevant to time scales shorter than 30 or so years.

What is particularly strange about the joint strategy recommendation is that it comes from researchers who not only know that the butterfly effect makes this recommendation almost impossible to implement, but they also have argued for the existence of something they call the "hawkmoth

---

[12] I add the parenthetical prefix because the UKCIP is a much discussed case of a climate modeling project that almost certainly oversteps the epistemic limitations of the craft. (See Frigg, Smith, and Stainforth 2015)

[13] (See L. A. Smith 2002 and Daron and Stainforth 2013)

effect" which, if their claims about it were true, would make the recommendation doubly impossible.

The "hawkmoth effect" is a term that was introduced by Erica Thompson (2013), but was motivated by work that originated with Leonard Smith (2002). It has been much ballyhooed in recent philosophical work spearheaded by Smith and Roman Frigg. The idea of the "hawkmoth effect" is that predictions made with non-linear models are just as much as sensitive to small errors in the correct structure of a model as they are to small errors in initial conditions, and that therefore, PICEF forecast methods, and thus non-linear models, are unlikely to be useful in generating "decision relevant" probabilities.

My own view is that most of these claims are misleading. First, I am skeptical that the term the "hawkmoth effect" is a helpful notion. The group usually introduces it in one of two ways: either using an example based on the logistic equation (see above) which they sometimes call the "demon's apprentice" example, or by appealing certain results in the topology of dynamical systems. Both their discussions of both the example and the results are somewhat misleading. And second, even if the idea of the hawkmoth effect were helpful for understanding the limitations of PICEF, it would still not be, as it purports to be, relevant to climate projects like the IPCC and the UKCIP. That is because neither of these involve PICEF at all. As we have emphasized, they appeal overwhelmingly to forcing experiments. But for whatever reason, the idea of it has made a very large encroachment into the philosophical literature on climate science, with four or five[14] (depending on how you count) philosophical papers published in what is still a young and underdeveloped field (philosophy of climate science). Because of this, it is worth spending a little bit of time explaining what the hawkmoth effect is supposed to be and why philosophers interested in climate science should not pay attention to it.

## 5.6 The Hawkmoth Effect

Despite the number of papers on the hawkmoth effect, nowhere is it precisely defined. But whatever it is, it is a very bad thing. It is a "poison pill'" that "pulls the rug from underneath many modeling endeavors'". It undermines the quantitative predictive power of almost all non-linear

---

[14] (Frigg, Smith and Stainforth 2013, Frigg et al. 2014, Bradley et al. 2014, Frigg, Smith and Stainforth 2015)

models and makes them incapable of producing "decision-relevant predictions" and "decision-relevant probabilities."

The LSE group[15] has been primarily concerned with climate science, and in particular with the use of climate models to produce probabilities of future climate outcomes from initial conditions. Their conclusions are highly skeptical, arguing that the only trustworthy source of quantitative knowledge concerning the climate system comes from non-dynamical equilibrium models.

To understand what the hawkmoth effect is supposed to be, it will help to note that it is conceptualized on analogy with the butterfly effect. Indeed, nothing captures the intuitive idea of a hawkmoth effect better than the following equation:

Butterfly Effect:   Initial Conditions

::

Hawkmoth Effect:   Model Structure

In other words, the hawkmoth effect is supposed to do to small errors that we might have in what we take to be the correct model of a system what the butterfly effect does to small errors in what we take to be the correct initial conditions of a system: quickly and systematically blow them out of proportion until they undermine our ability to make useful predictions. Why, though, does the LSE group think there is such a thing? Why is the butterfly effect so famous and why, as they sometimes put it, does the "less media-friendly hawkmoth [] not get as much attention as its celebrated butterfly cousin?"[16] Worse still, why do they think that most non-linear models display these features? Or that the burden of proof is on anyone who uses non-linear models to show that the models *don't* have this feature?

I can at best only answer the first question. They have two reasons. The first has to do with a set of mathematical results associated with the topological concept of "structural stability." Indeed the closest thing one gets to a definition of the hawkmoth effect is in (Thompson 2013) where she more or less identifies it with the absence of structural stability.[17] The second

---

[15] Most of the authors of these papers have or have had some affiliation with the London School of Economics, so I use the label here for convenience.

[16] (Thompson and Smith 2017)

[17] "I propose that the term "Hawkmoth Effect" should be used to refer to structural instability of complex systems." (Thompson 2013, p. 213.) All we get in the philosophical papers is "Thompson (2013) introduced this term in analogy to the butterfly effect. The term also emphasizes that SME yields a worse epistemic position than SDIC." (Frigg et al. 2014, p. 39).

reason has to do with an example they demonstrate using the logistic equation that they call "Laplace's demon's apprentices."

As several co-authors and I have argued elsewhere, there are two confusions underlying all of this. One source of confusion, as William Goodwin and I argued in Winsberg and Goodwin (2016), is that they do not adequately appreciate the differences between PICEF and projections based on forcing experiments (that we just discussed in section 5.4). Recall that forcing experiments do not depend at all on the actual initial conditions that obtain in the world. They depend only on having a reasonably good sample of initial conditions. Much of the work that they do is to try to show that small model errors can ruin PICEF forecasts, but never acknowledge the minor role that PICEF plays in climate science. A deeper source of confusion, perhaps, is that they misunderstand or misrepresent the nature of structural stability, which is a topological, not a metrical, notion, and which does not have an interesting complement in the form of an "instability" – let alone one that would correspond to structural mixing or sensitive dependence on model structure. Absence of structural stability is much weaker, in some respects, than an analogy to the butterfly effect would suggest, and simply completely disanalogous in others. They also misunderstand its connection to the logistic equation (which is not a diffeomorphism and hence does not even fall into the category of being structurally stable or not). Finally, they mispresent what their logistic equation example can demonstrate – by employing a faulty notion of two models being "nearby." All of this is argued for by Alejandro Navas, Lukas Nabergall and myself (forthcoming). And all of this is a very good thing, because despite what they claim in some of their papers, as we argue in Winsberg and Goodwin (2016) and in Navas, Nabergall and Winsberg (forthcoming), it would be devastating for climate science, for attribution claims, and for any policy making recommendations based on climate science if any of their claims were *true*. Interested readers should consult some of the referenced papers for more details, and the appendix at the end of this book.

## 5.7 Conclusion

The atmosphere is almost certainly a chaotic system. Among other things, this means that nearby initial states of the system diverge in the system's state space very quickly. This, in turn, means that making ordinary weather predictions about the atmosphere becomes nearly impossible for longer periods of time. There are a number of ways of mitigating this problem. One of them is to use Probabilistic Initial Condition Ensemble Forecasting (PICEF). But the primary way that climate science deals with the chaotic nature of the atmosphere is by mostly making *projections* rather than *predictions*. Projections are supported by "forcing experiments" where we investigate how the statistical and qualitative properties of a system respond to a change in a model parameter value. These are immune to worries about the "butterfly effect" that we associate with chaotic systems. Worries, furthermore, about a model-structure analog of the butterfly effect, which has sometimes been called the "hawkmoth effect" are over-blown. This is both because there is no close model-structure analog of the butterfly effect, and because even if there were, it would not be likely to affect climate *projections*.

## Suggestions for Further Reading

Kellert, Stephen. 1993. *In the Wake of Chaos*. The earliest philosophical work on chaos theory. Includes a very readable introduction to the main ideas of chaos theory and a nice discussion of some of its implications for such philosophical topics as unpredictability, scientific explanation, and feminist philosophy of science.

Smith, Peter. 1998. Explaining Chaos (especially Chapters 1 and 4). A more mathematically rigorous presentation of chaos theory that is still targeted at philosophers. Contains a useful presentation of many of the crucial definitions.

Slingo, Julia and Tim Palmer. 2011. "Uncertainty in Weather and Climate Prediction," *Philosophical Transactions of the Royal Society*. A beautiful presentation of the role of uncertainty in climate and weather. This is the background material to all of section 5.4. Contains excellent illustrations.

Frigg, R., S. Bradley, H. Du and L. Smith. 2014. "Laplace's Demon and the Adventures of his Apprentices," *Philosophy of Science*. One of the famous "hawkmoth" papers. Argues that the phenomenon of structural stability and its complement makes non-linear models unsuitable for numerical prediction and the generations of decision-relevant probabilities.

Winsberg, Eric and William M. Goodwin. 2016. "The Adventures of Climate Science in the Sweet Land of Idle Dreams," *Studies in History and Philosophy of Modern Physics*. A reply to Frigg et al. Argues that the principle argument they employ is misdirected at climate science, which is immune to the kind of initial condition uncertainty they are concerned with.

Navas, Alejandro, Lukas Nabergall and Eric Winsberg. (Forthcoming.) "On the Proper Classification of Lepidoptera: does the absence of structural stability produce a 'hawkmoth effect' that undermines climate modeling?" Another reply to Frigg et al, this time arguing that the very idea of a "hawkmoth effect" is highly misleading.

Stewart, I. 2011. "Sources of Uncertainty in Deterministic Dynamics: an informal overview," *Philosophical Transactions of the Royal Society, A.* Somewhat similar to Slingo and Palmer, but taking a wider view of uncertainty than just the butterfly effect by looking at the various topological sources of uncertainty in dynamical systems.

# 6

# *Probability*

## 6.1 Introduction

In this chapter, we will be concerned with the question of how to interpret the probabilities in climate science. Let's start with the claim that there *are* probabilities in climate science. The fact that climate science concerns itself with probabilities is obvious if we consider the following passages taken from the Fifth Assessment Report (AR5), Working Group I, of the IPCC:

> It is *very likely* that regional trends have enhanced the mean geographical contrasts in sea surface salinity since the 1950s: saline surface waters in the evaporation-dominated mid-latitudes have become more saline, while relatively fresh surface waters in rainfall-dominated tropical and polar regions have become fresher. (*high confidence*) (p. 40)

> It is *virtually certain* that the increased storage of carbon by the ocean will increase acidification in the future, continuing the observed trends of the past decades. (p. 469)

> The water vapour/lapse rate, albedo and cloud feedbacks are the principle determinants of equilibrium climate sensitivity. All of these feedbacks are assessed to be positive, but with different levels of likelihood assigned ranging from *likely* to *extremely likely*. Therefore, there is *high confidence* that the net feedback is positive and the black body response of the climate to a forcing will therefore be amplified. (p. 82)

> The net radiative feedback due to all cloud types is *likely* positive. (p. 82) (*Confidence* in this claim is *low*)

> Equilibrium climate sensitivity (ECS) is *likely* in the range 1.5°C to 4.5°C with *high confidence*. ECS is positive, *extremely unlikely* less than 1°C (*high confidence*) and *very unlikely* greater than 6°C (*medium confidence*). (pp. 83–84)[1]

Notice the emphasized expressions. If we look carefully, we can see that there are expressions of likelihood and expression of confidence. Usually,

---

[1] All quotations from IPCC, AR5, WG1

Table 6.1 *Interpretations of probability terms in the IPCC reports*

| Term | Likelihood of outcome |
|------|----------------------|
| Virtually certain | >99% probability |
| Extremely likely | >95% probability |
| Very likely | >90% probability |
| Likely | >66% probability |
| More likely than not | >50% probability |
| About as likely as not | 33 to 66% probability |
| Unlikely | <33% probability |
| Very unlikely | <10%probability |
| Extremely unlikely | <5% probability |
| Exceptionally unlikely | <1% probability |

expressions of confidence are applied *to* expressions of likelihood: e.g. there is high confidence in the claim *that* it is very likely *that* e.g. the increased storage of carbon by the ocean will increase acidification in the future. Let's start with the first kind of expression, which is the simpler of the two, and we'll come back later to a discussion of the phrases involving the word "confidence" in the next chapter. In the AR4, the IPCC provides the following chart[2] to explain the first kind of phrase:

## 6.2 Interpretations of Probability for Climate Science

It should come as no surprise that the IPCC should want to attach probabilities, or at least ranges of probabilities, to various scientific hypotheses about the climate – especially about its future development. Scientists (and we who follow their activities) have varying degrees of belief in most of the scientific hypotheses we entertain. Some of our scientific beliefs we hold very firmly (that the earth is round) and some we think are considerably more likely to be true than not, but we are far less sure of them. (For example, most, but not all, astrophysicists believe that the predominant source of the gravitational force that holds rotating galaxies together is cold dark matter.) Some scientific beliefs are much closer to being mere speculations (e.g. our universe comes from a big bang which is just one

---

[2] See IPCC, AR4, WG1, 1.6, Box 1.1. (See also Mastrandrea et al. 2010, 3.) In fact I would argue that the chart makes more sense (in conjunction with everything else the IPCC says) if we understand the bins as exclusive, rather than overlapping. In other words, "very likely" is actually 90%<pr<95% rather than simply pr>90%.

emergent bubble in a multiverse with an infinite number of bubbles). These attitudes to our scientific hypotheses are no different from the ones we have regarding our ordinary, everyday knowledge of the world: I'm virtually certain that I have two hands. I believe Jerry Garcia had only nine fingers, but I am considerably less certain that this claim is true.

One additional assumption I can make about the nature of people's attitudes about the scientific hypotheses they entertain is that I can represent the strength of these degrees of belief by assigning numbers to them between zero and one (or between 0 and 100 percent) in proportion to how strongly they believe them. If the numbers I assign to a variety of beliefs meet a minimum set of constraints, then we can call these numbers probabilities.[3] As the chart above makes pretty clear, probabilities play an important role in climate science and climate policy guidance. Given the important role of probabilities in both weather forecasting and climate science, it behooves us to get a bit more clear about what probabilistic statements of these kinds actually mean in each of these two disciplines.

Given any statement, S, we can assign a probability to that statement: the statement that the earth is round, that the next flip of this coin will be heads, that it will be cloudy tomorrow, that I have two hands, that we live in a multiverse with an infinite number of bubbles, or that ECS is in the range 1.5°C to 4.5°C. What it means to say that a number that I assign to a statement is a *probability* is just that the number in question, along with all the other numbers I assign to my other beliefs, obey the following axioms of probability, which are known as Kolmogorov's Axioms (Box 6.1).

The fact that I have assigned numbers to the statements I entertain, and that those numbers meet the requirements of being probabilities (the three axioms above) doesn't, however, tell me much about what probabilities *are*. What does it mean to say that the probability of S is pr(S)? Alternatively, what are the truth conditions of the probability of S being pr(S)? Philosophers, of course, have had a lot to say about this, and most of it is fairly controversial. But *most* philosophers agree on at least this much: probabilities come in at least two varieties, each of which are called by a variety of different names. We can start by using the following pair of expressions: "subjective probabilities" and "objective probabilities."

---

[3] In fact, there are some widely accepted arguments (the most famous of which is the Dutch Book argument) that any person's degrees of belief over all the propositions she entertains must meet those constraints if she is to count as rational.

> ### Box 6.1 Kolmogorov's Axioms of Probability
>
> If **S** is a set of statements that are closed under truth-functional combination, and pr(S), pr(T), etc., are numbers associated with those statements, we can call those numbers probabilities, if and only if,
>
> 1) $0 \leq pr(S) \leq 1$ for all $S \in$ **S**
> 2) If S is certain, then $pr(S) = 1$
> 3) If S and T are incompatible, then $pr(S \vee T) = pr(S) + pr(T)$
>
> A similar axiomatization can be given where **S** is a space of events, and the truth-functional combinations are replaced with set-theoretic operations in the ordinary way.

## 6.3 Chances and Credences

"Subjective probabilities" is the most common name for the first kind of probability (they are also often called "credences" or "personal probabilities" or "degrees of belief"), but the name is somewhat unfortunate, because the word "subjective" is fairly ambiguous. It is used to mean a variety of different things. Sometimes it means the opposite of "objective," in the sense of "a determinate matter of fact." ("Your opinion of this book is merely subjective!") And sometimes it means the opposite of "objective" in the sense of "genuinely real/part of the furniture of the world."

When we speak of "subjective probabilities," we mean "subjective" in the second sense of the word – in the sense that they are not mind-independent things like tables and chairs. Subjective probabilities are taken by most philosophers to be dependent on the mental states of individual agents. The two notions of subjectivity, admittedly, are not unrelated. But it is a further question whether subjective probabilities are *also subjective* in the sense that we needn't take any particular value of them to be objectively correct – that there is no determinately correct value that they take. Many philosophers think that subjective probabilities are also *subjective* in the *second sense*. But that is a substantive thesis, not a matter of a definition we should adopt. Many "Bayesians," as it happens, argue that the only rational constraint on subjective credences is that they must obey the axioms of probability: as long as your credences obey the axioms, they are just as good as anybody else's. But just obeying the axioms is a very weak demand. So not everyone has this liberal of an attitude. Some "objective Bayesians" believe there are objectively correct credences to hold. And of course, you might have a view somewhere in between – that objective correctness places *some* further constraint on degrees of belief, but doesn't nail it down entirely. For these reasons, the expressions "credence" or "personal probability" are somewhat more apt.

So credences are degrees of belief. They are psychological characteristics of individual agents. But how do we tell what credences some agent holds? How do we tell what degrees of belief they have in various propositions? One way is just to ask them. In other words, credences can be numbers that experts (and others) can deliberately and self-consciously assign to a set of hypotheses that they entertain as a way of communicating the strength of their beliefs in those hypotheses to others. Unfortunately, that doesn't tell us much about what it means to have such and such a credence. Another answer is to say that the criterion for some rational agent having degree of belief, or credence, P, in some event, x, under circumstances, C, is that the agent is willing to bet on x, in circumstances C, in accord with probability P. If an agent values dollar bills four times as much as she values doughnuts, and she is willing to bet doughnuts for dollars that event x will happen under circumstances C, then we can say that she and her counterpart's credence in x given C, is .2. On this common view, the rational agent's betting preferences are not (merely) *evidence* of what her credences are, they are also *constitutive* of them.

We'll say a little bit more in what follows about what to think about these issues when talking about climate hypotheses. For now, I just want to note that there are a variety of ways of thinking of credences or personal probabilities, and they needn't be cashed out in such purely psychologistic terms (where dispositions to bet are constitutive of credences). In Chapter 8 we will discuss other conceptions of credences according to which credences are things that agents choose, rather than things that they just have, and we will give reasons to think that maybe those conceptions make better sense of the probabilities in the IPCC. For now it will suffice to be clear that credences are connected to epistemic agents, and not just to the mind-independent systems and events they describe.

Objective probabilities (which are sometimes called "objective chances," or just "chances"), on the other hand, are meant to be features of the mind-independent world. They are meant to be properties, directly, of the events they describe or the systems that give rise to them. Some philosophers (a small minority?) don't believe in objective chances. But one compelling reason to think that we need objective chances is that some of our most important scientific theories make use of probabilities: quantum mechanics (QM) and statistical mechanics (SM) at least standardly appear to do so. Scientific theories, we would like to think, tell us about a mind-independent world – not about anyone's private mental states. And so it seems like the probabilities that QM and SM report are objective chances that are features of the world.

Many philosophers believe that the only objective chances that there are in the world are the ones, if there happen to be any, which follow from fundamental physical laws that themselves involve a stochastic element. But this still seems wrong. We standardly say that roulette wheels and other classical games of chance exhibit objective chances, and our confidence in that claim doesn't seem to depend in any way on our having unlocked the mysteries of quantum mechanics. So I will be assuming here that it is perfectly possible for chance systems to exhibit (non-trivial) objective chances even if they are made up out of stuff that obeys deterministic laws.[4] Indeed, I would suggest that on the right account of objective chance it should turn out that there are true chance statements about the *weather*. Nevertheless, I still think it will turn out to be very unlikely that there are true chance statements about the *climate*.

It is quite a bit easier to spell out what personal probabilities are than it is to spell out what chances are. Chances are notoriously difficult to characterize. Some philosophers think that chances are not only mind-independent entities, but that they are irreducibly fundamental features of the world. Others believe that chances supervene on non-chance involving features of the world. Both of these views, unfortunately, share difficulties. But one thing that most people agree on is that chances are in some way connected to frequencies of outcomes (it is the direction of dependence that is most controversial). Famously (or maybe infamously), chances cannot be literally identified with frequencies, but they do seem to have a very tight connection with them.

One reasonably clear fact that I will appeal to here is that you get objective chances associated with "chance systems" that display stable enough patterns of frequencies. We can make true claims about many systems that involve objective chances: roulette wheels, dice, flipped coins, shuffled cards, medical patients (An N-year-old male with stage 3 lung cancer has an X percent chance of survival), and many others. The reason for this has to do with the stable patterns of frequencies we find when we study these kinds of system – though the connection between the two is not always simple or direct. Spelling out the exact connection gives rise to difficulties, because we want to be able to maintain the possibility of singular events having chances, and we want to maintain the possibility of

---

[4] In fact, I will be closely following a defense of this claim found in Hoefer (2007)

chances differing from frequencies (they can sometimes, for example, be irrational numbers). Furthermore, we would like to be able to say that the chance of a suitably symmetrical 145-sided dice landing on 25 is 1/145 even if no such dice has been made. These are among the reasons why chances can't themselves be frequencies. Nevertheless, chances and frequencies go hand in hand in some sort of way.

## 6.4 … And How to Tell Them Apart

How do we tell, about a certain kind of statement, S, if there are true claims of the form, "The objective chance of S is pr(S)," where pr(S) is neither zero nor one? This is not a trivial question. One way to know for sure that the answer is "yes" is if the statement S is about a system that is governed by non-deterministic laws of evolution. Thus, if it turns out to be true that the correct version of quantum mechanics is one in which there is a fundamental law of nature that causes the wave function to collapse in accord with certain fundamental probabilities, then it will definitely be the case that some statements of the form "The chance of the photon passing through the polarizer is .5" will be true. That will obviously be the case even for polarizers that we never build. So being a probability derivable from a non-deterministic law of evolution is a sufficient condition for being a chance. But this is much too strong, for reasons we have discussed, to be a necessary condition – there is good reason to think that a system can exhibit objective chances without its underlying components obeying stochastic laws. We would very much like it to come out that it is a true statement to say "The objective chance of a roulette wheel landing on 5 is 1/38" even if the world were governed by a deterministic micro-physics.

Consider the following pair of examples. Mary is in the casino and we observe her make two different bets. She is willing to bet at 37 to 1 odds that the roulette wheel (with 38 slots) will land on "5," and she is willing to bet $5 to win a $25 pot on the proposition that her opponent at the poker table is bluffing. In both cases, she has a credence that is reflected in her willingness to bet. But in one case that credence (directly) reflects her belief in an objective chance and in the other it doesn't.

In the first case, she believes that the roulette wheel is a chance setup and that the probability of her getting "5" is 1/38. Why does she believe this? One factor is that she has seen the wheel land on each slot about an equal number of times. But that is only one factor. She would probably be willing to bet even if she had never observed a 38 slotted wheel before (many casinos have 37 slotted wheels).

So what are the other factors? First, there is the physical symmetry of the wheel. But equally importantly, there are many elements that randomize the toss of the ball and the spin of the wheel: the spinning of the wheel and the behavior of the croupier randomizes the initial entry-point of the ball, the way the ball bounces off the various ridges and bumps on the wheel makes its behavior chaotic, and the various outside perturbations on the system are also random, and further add to the randomization of the system. Mary has a credence of 1/38, in the proposition that the ball will land on "5," but it is not a "mere" credence. The credence reflects her belief in an underlying chance.

In the second case, Mary might also believe there is a chance setup involved in the shuffling and distribution of the cards, but she will not want to set her credence exactly in accord with the chances associated with that system. In addition to knowing that the chance her opponent has drawn an inside straight, is, say, 1/13th, she also has additional information that affects her credence: the proposition that her opponent is bluffing, which does not come by way of impacting her assessment of the original chance of the card filling the inside straight. For example, she has the information that her opponent bet heavily at first, but is now backing off. She has the information that he is frequently touching his betting chips, etc. She has the information that he smiled when he drew his penultimate card, but not when he drew the last one. She modifies her degree of belief in the face of all this evidence. The fact that she is modifying her degree of belief in the proposition that her opponent is bluffing based on information that *does not affect her beliefs about what kinds of chance systems are involved* indicates that her betting behavior is based on credences that do not (exactly) reflect underlying chances.

In other words, there are two telltale signs that distinguish degrees of belief grounded in chances from those that are not.

1) *All* of Mary's reasoning about how to set her credence vis-à-vis the roulette wheel involves reasoning about observable stable frequencies in the world. It could be a direct route from frequencies to probabilities (if she's been studying the particular wheel herself and made a note of its outcomes), or (more likely) because of beliefs she has about the underlying processes that drive roulette wheels in general (spinning, bouncing, external perturbation, etc.) and the stable patterns of frequencies they exhibit.

Mary is setting her credence about her poker opponent using lots of other kinds of information – *information that is not relevant to her beliefs about the behavior of any sort of chance system.* She believes that touching your chips is a sign of nervousness; that not smiling when you get your card, if you smiled on the last card, is a sign that you didn't get what you wanted.

2) Mary's credence about how the roulette wheel is going to land is extremely resilient to new information. Nothing about how the croupier is behaving is going to affect her credence. The only evidence that will affect Mary's credence is evidence about what the relevant behavior of chance systems is – evidence she is unlikely to find in a casino. But Mary's credences about her poker opponent are subject to constant updating. This kind of resiliency is one sign of a credence based on a belief in an objective chance.

## 6.5 Probabilities in Weather and Climate

### *i) Weather*

Weather systems sometimes behave like chance setups in the same way that roulette wheels do. There are various macroscopically observable states in which the atmosphere can be. And there are various processes which can randomize the underlying microscopic components in virtue of which that macroscopic state can be realized, 30 percent of which lead to rain, such that it makes sense to say that the objective chance of rain is 30 percent. Weather forecasters sometimes set (and announce) their credences in accord with these sorts of chances.[5] So the probabilities in weather forecasting are sometimes chances.

One telltale sign that an announced 30 percent probability of rain is an objective chance is that it will turn out to be resilient to a wide range of admissible evidence. Forecasters could easily acquire new evidence that the atmosphere was no longer in the chance setup state that they thought it was. More radically, they could learn things that would make them change their minds about what the objective chances are for such setups. But this is quite different from what you and I do (or at least I do this rather often) when we set our own credence in the proposition that it will rain. If I see my neighbor walking out the door with an umbrella, then I raise my credence. If I then see my neighbor putting the umbrella in what looks like

---

[5] Or at least in principle they could. In fact the procedures for assigning probabilities to weather forecasts are probably quite a bit more complicated than all of the above makes it out to be, but this is not a book about weather forecasting. I merely want to make the point that, unlike in the climate case, there are genuine chances regarding the weather, and a science of weather forecasting could in principle discover them. Whether or not it actually does is a further question.

a pile of stuff bound for the goodwill donation box, I lower my credence. Credences that reflect underlying chances are more resilient than this.

Notice, by the way, that at least for fine-grained forecasts, for example whether there will be rain on a particular day in a particular zip code, even probabilistic forecasts of weather are pretty useless if we try to make them for too far in the future. This could either be because there are no forecast-worthy objective chances for, say, whether or not it will rain in Tampa 20 days from now[6], or it could be that there are forecasting chances but we are ignorant of them. Let's use Figure 5.1a as a cartoon of what happens in weather forecasting. Think of the right-hand lobe of the Lorenz attractor as corresponding to it raining in Tampa, and the left-hand lobe as corresponding to it not raining in Tampa. Now think of the circle in the top-left corner as representing the set of microstates consistent with the presently observable macrostate of the atmosphere. A few points are worth making.

First, suppose that I make a forecast by putting a uniform probability measure on that top-left circle. I get a resulting probability distribution that is uniform over the subsequent patches in the diagrams. What has to be the case for these to count as genuine chances, as opposed to merely underwriting my personal degrees of belief? It has to be the case that the original probability distribution is not just my degree of belief, but that it in turn represents objective chances. And for that to be the case, it has to be that the properties of the weather system genuinely randomize the initial conditions over that patch. The patch would have to be characterized by a set of microstates that were mixed by internal variability. Objective chances have to be tied to stable frequencies, and in this case the stable frequencies are the ones associated with the way the chaotic atmosphere randomizes its initial conditions.

Second, suppose I want to know the probability of rain in Tampa in 60 units of time (where a unit of time corresponds to the evolution of the system for one solid white arrow in the figure). It will probably turn out that after 60 units of time, my patch has spread out all over the attractor (the way things look in Figure 5.1c). That might correspond to a 50 percent chance of rain in Tampa, but that's not really an interesting weather forecast – it is not a "forecast-worthy chance" in the sense that it is not

---

[6] I say no "forecast-worthy" objective chances because there might still be the chance of rain associated with simply being a summer day in Tampa, but this is not a chance conditional on an initial condition, and so it is not much of a weather forecast.

really conditional on the present state. It is just the chance of rain in Tampa on any old summer day.

### ii) Climate

There are, therefore, true claims about objective chances in weather forecasting. *But we probably don't have access to such true claims about the climate.* It follows from this that *if* present-day climate science is in the business of reporting probabilities, then we have to make sense of those probabilities as credences.

Why think there are no objective chances in climate science? For one, we don't know of any climate variables that exhibit, or are appropriately linked to, the kind of stable frequencies that usually underwrite objective chances. Precisely because weather reflects, at least in large part, the internal variability of a chaotic system, weather events are genuinely chancy events – and sometimes we can come to know what those chances are. But climate variables are precisely those that average over internal variability. There doesn't appear to be any kind of system-that-exhibits-stable-frequencies that gives rise to *climate* outcomes. A clear symptom of this is that there are no probabilistic hypotheses in climate science that we expect to be sufficiently *resilient* in the face of new evidence in order to count as reflecting objective chances. We can expect climate scientists to adjust the probabilities that they assign to climate hypotheses in the face of a very wide variety of possible new evidence – and not only in the face of the kind of evidence that makes them revise their views about the kinds of stable frequencies that chance systems exhibit.

To be clear: I am not arguing that we can know a priori that there are no chances associated with the climate. Here is one way that there could be. Suppose that the earth is sitting near the tipping point between two different futures. In one such future, we double the quantity of carbon in the atmosphere and the earth responds with a 2°C warming. In the other future, we double it and the earth responds with a 4°C warming. And suppose that the difference between these two futures, as a matter of physics and geodynamics and the like, was that the former happened when the earth is in 40 percent of the microstates compatible with its presently observable state, and the latter happened in the other 60 percent, and it was a feature of the earth's chaotic dynamics that there was, in the current climate regime in which we live, a stable set of frequencies that underwrote the claim that there was a uniform probability distribution (as a matter of objective chance) over that set of microstates. It would then be the case that there was a 40 percent *chance* of the earth's climate sensitivity being 2 °C, and a 60 percent *chance* of it being 4 °C. But of course all of that is science fiction (as far as we know). We have no good reasons to think that any climate outcomes are conditional on macrostates that are characterized by a set of microstates that are mixed by internal variability. So we have no particularly good reasons to believe such chances exist and we certainly don't have any good reason to think we know any of them.

So the uncertainties that we deal with in climate science, the ones faced by climate experts and reported by the IPCC, are *epistemic* uncertainties. They are the kind we could reduce, in principle, with better models, etc. Hence we report them by reporting subjective, or personal probabilities: *credences*.

The probabilities in the above chart, used by the IPCC, therefore, are best thought of as credences. But that's a little bit too fast. Above we said that many philosophers take credences to be "psychological characteristics of individual agents." The IPCC is not an individual agent. This complicates matters, since it means we have to interpret the IPCC's reported probabilities somewhat differently than philosophers usually interpret the subjective credences of individual agents.

We have several choices here. We could say that when the IPCC says "It is *very likely* that regions of the ocean with high salinity where evaporation dominates have become more saline," what they mean is that everyone, or almost everyone, or most people (more on this point later) involved in the panel has a degree of belief higher than 90 percent in the proposition that regions of the ocean with high salinity where evaporation dominates have become more saline. Or we could say that they mean that the consensus of the panel is to recommend that the readers of their report adopt a degree of belief of at least 90 percent in the proposition that regions of the ocean with high salinity where evaporation dominates have become more saline. Or we could follow the intuitions of objective Bayesians and say that the IPCC is reporting the objectively correct probability to hold in light of the presently available evidence (but then we would need to find a logic of confirmation according to which this was the case). Or we might find some other interpretation that is somewhere in the neighborhood of these three different claims. I will say more in due course about how you can make sense of a report of the mental states of a *panel of* experts. Inter alia, we will see that one additional advantage of this approach is that it will help us to better understand the second kind of language in the quotations at the top of the chapter – the language about degrees *of confidence*.

At this point, we need to address an important concern that some readers might have with respect to the role of credences in climate science. I can easily imagine the reader who is thinking the following: "If the world is such that our best fundamental theories have to make irreducible appeal to probabilities, then I can certainly see why scientists should use and report objective chances. After all, on that assumption, that would be the only way for them to accurately describe the mind-independent world. And even if our best theories are deterministic, but there are stable chance setups in the world that produce stable frequencies, then I can see why it would be the job of scientists to summarize those for us using chances. But if there is no room for objective chances in climate science, why should climate scientists be trafficking in descriptions of their own mental states?"

We usually think, in other words, that the job of scientists is to tell us about the world, not their own minds! Perhaps, in fact, the reader I have in mind would express their concern in the following sort of way: with all the controversy surrounding climate science, shouldn't climate scientists stay away from giving *subjective* reports?

Let's tackle the last worry head-on: we need to be careful here not to confuse, as Rougier and Crucifix (2018) have warned, the Mertonian norm of scientific *disinterestedness* with the kind of objectivity that is the opposite of the "subjectivity" in "subjective credences." Reporting credences does not conflict with the Mertonian norm – it is not a failure of disinterestedness. One can perfectly well, at least in principle, reach a careful judgment about what personal degree of belief to adopt in a relevantly disinterested manner. When Mary was setting her credence about the proposition that her opponent was bluffing, the only interest she had was in maximizing her expected return on her bets. Her credence, therefore, was subjective, but relevantly disinterested. To see the point, consider the contrast with the case in which her opponent's friend (who will perhaps share in her opponent's winnings) is advising her on whether to raise or fold. *That* person's advice assignment of a probability might not be disinterested.

## 6.6 Conclusion

If climate science and climate models are going to be tools that policy makers can use, then their users and practitioners need to report probabilities, and the Lyndon B. Johnson principle applies ("I'm the only president you've got"). Subjective credences are the only probabilities that make any sense in climate science.

Why believe that climate science as a guide to policy needs to report probabilities? Because probabilities are the very best way for communicating knowledge under uncertainty from *experts* to *policy makers* in a way that does the best we can to preserve the value-free ideal of science. Imagine that you are sitting at the poker table in the above example but don't consider yourself to be a very skilled poker player. Lucky for you, Mary is sitting next to you, and is willing to give you advice. In such a scenario, you basically have two choices: you can either have her decide for you whether to call or fold, or you can have her tell you what her credence is in the proposition that you will win or lose if you call. The first choice might be satisfactory, but what if Mary doesn't know your financial situation. She doesn't know what the impact on your life will be if you win or lose. Maybe she doesn't know that if you lose this bet, you won't be able to afford shoes for your children. Or conversely, that *unless* you win, you won't be able to afford the shoes anyway. In that situation, you are much better off having her tell you the probabilities and deciding for yourself whether to bet.

The situation in climate science is analogous. Experts in climate science have knowledge about the climate. In one sense, therefore, they are the people who ought be considered best situated to make decisions about what we ought to do in matters related to climate. But in another sense, they are not.

Consider various possible climate adaptation strategies. Here is one: one of the many possible dangers of regional climate changes are *glacial lake outburst floods*. These occur when the dam (consisting of glacier ice and a terminal moraine) containing a glacial lake fails. Should a local community threatened by a possible flood replace the terminal moraine with a concrete dam? The answer to this question surely depends on the likelihood of the glacier melting and the existing (natural) dam bursting. And so we might think that the people best positioned to answer the question are the climate scientists with the most expertise about the future of the local regional climate. On the other hand, the answer also surely depends on the cost of building the dam, and the likely damage that would ensue if the dam were to break. Just as much, it might depend on how the relevant stakeholders weigh the present costs against the future damages. And so while on the one hand we would like the people making the decision to have the most expertise possible, we also, on the other hand, want the decision to be made by people who *represent our interests*, whoever *we* might be. Making decisions about, for example, climate adaptation strategies, therefore, requires a mixture of the relevant expertise and the capacity to *represent the values*[7] of the people on whose behalf one is making the decision. But there is rarely any single group of people who obviously possess both of these properties. And even if there were, it would just be, in effect, a coincidence. The expert qua expert does not represent our values, and the policy maker qua representative of the public will is not an expert of the relevant kind (or is not an expert about the climate).

So climate science in the service of policy making *should* report probabilities if it wants to try to maintain the value-free ideal, and the IPCC does so. And the best way to understand those probabilities is as a summary of the personal degrees of belief of a panel of relevant experts. This still leaves at least three questions we could ask about the IPCC's probabilities. We will turn to them in the next chapter.

## Suggestions for Further Reading

IPCC, AR4, WG1, 1.6. The canonical statement of what uncertainty terms mean in all IPCC reports. This is still more or less the standard, though small modifications were made in the next entry:

Mastrandrea, Michael et al. 2010. "Guidance Note for Lead Authors of the IPCC Fifth Assessment Report on Consistent Treatment of Uncertainties." A new document for the AR5 on the treatment of uncertainties, which also includes a treatment of levels of "confidence," which we will discuss in more detail in Chapter 7.

Hájek, Alan. 2003. "Interpretations of Probability," *The Stanford Encyclopedia of Philosophy*. An excellent appraisal of a century's work on the interpretation of probability by one of its most prolific recent contributors. Contains an extensive bibliography.

Hoefer, Carl. 2007. "The Third Way on Objective Probability: A Sceptic's Guide to Objective Chance," *Mind*. A compelling defense of the claim that there can exist objective chances in a deterministic world.

Rougier, Jonathon and Michael Crucifix. 2018. "Uncertainty in Climate Science and Climate Policy." An excellent discussion of subjective probability in climate science and some of the difficulties associated with assigning them, written by a statistician and a climate scientist writing (in part) for a philosophical audience.

---

[7] Of course one might have worries about whether elected representatives generally represent the values of their constituents, but that is a subject for a different sort of book.

# 7

# *Confidence*

## 7.1 Introduction

In the last chapter we reached the conclusion that climate science in the service of policy making should attach probabilities to the hypotheses it evaluates. It must, if it wants to do the best that it can to maintain the value-free ideal of science. The IPCC does, moreover, report probabilities. And those probabilities can only be credences, because there are no objective chances in climate science. But credences are mind-dependent, agent-relative entities, and the IPCC is not a single agent. So this leaves at least three questions we could ask about the IPCC's probabilities:

1) What are the sources of the uncertainty with regard to climate projections such that they need to be reported as probabilities?
2) How do climate scientists arrive at the probabilities they attach to their projections? (How do they do it *in fact*? And how *should* they best do it?)
3) What's the right way to understand a report of credences that is associated with a group of experts, rather than a single agent?

We also, in the last chapter, set aside the question of how to understand the second-order qualifications that the IPCC attaches to some of the hypotheses it evaluates: those that express degrees of *confidence* in the first-order *probabilistic* claims. As we will see, this question is closely bound up with the third question, above, and also with the fact that climate science relies on a variety of sources of information to support its hypotheses.

## 7.2 Sources of Uncertainty

We'll start by talking about the sources of uncertainty. This is a list of items that are often mentioned as sources of uncertainty in climate modeling and climate projections:

a) Model structure
b) Numerical approximation
c) Parameters
d) Internal variability
e) Real data (observations)
f) Boundary values
g) Economic scenario.

Let's look at these one by one:

### a) Model Structure

Almost by definition, a model differs from the real-world system it tries to model. In the case of climate, we know, as a matter of fact, that even our best models leave out many of the components and interactions that are part of the real system we try to model. Insofar as they leave out these factors and interactions, no matter how much we tried to otherwise perfect our models, their predictions would still differ from reality to some degree.

### b) Numerical Approximation

Even if we had the perfect model structure, with every single component and every single interaction between those components included in the model, the differential equations we would end up with would not be amenable to analytic solution. A perfect model in terms of differential equations (a mathematical description of the rates of change in the system) would not give rise to a perfect mathematical description of the system's actual behavior. We would still have to use a computer to approximately "solve" those equations step-by-step, and that introduces errors. Conceptually, numerical approximation is a source of uncertainty, and it is distinct from model structure uncertainty.

In reality, however, it is somewhat misleading to separate numerical approximation from model structure as a source of error. Separating them makes it seem as though the choice of model is independent of computational exigencies – i.e. first there is an imperfect model chosen, and then it is merely approximately solved. But this is doubly false. If we could discretize climate models down to the millimeter, then we could probably come much closer to starting with something akin to the ideal model. For example: the modeling of cloud dynamics is generally thought to be a weak spot for climate models. But if the atmospheric dynamics were modeled down to the micrometer, we might not need a separate model of cloud dynamics. And conversely, given that we only discretize (typically) down to about 100km, we don't start with the "perfect model." The perfect model, after all, doesn't contain unrealistic parameterizations. Parameterizations, after all, are "imperfections" that are designed to counterbalance the effects of the coarse grid (see Chapter 4, section 4.4.d). Most parameterizations represent compromises between model-structure realism and computational exigencies[1].

In other words, one of the most important reasons that our models' structures are imperfect is because we are trying to mitigate the uncertainty caused by (very coarse!) numerical approximations – and there is no perfect way to do that. It makes no sense, for example, to ask of the GFDL2.x model how much error is contributed by the large grid size and how much by imperfect model structure. The two sources of error are inseparable. Many of the parameterizations that are part of the model's structure are there because the grid size is roughly what it is. Model structure – understood as the joint effect of model choice, discretization scheme and grid size, and parameterization scheme – should be considered as a single source of error.

### c) Parameter Values

Very closely related to the first two sources of uncertainty is uncertainty that comes from lack of knowledge of the best values to use as inputs to the various parameterizations. It is very closely connected to the first two sources both because whether or not a parameterization is needed in the first place depends on model structure and grid size, and because the best value of a parameter might also depend on grid size and on how other features of the model are implemented (see the section on modularity in Chapter 4). Notice, by the way, that I say "best" value rather than "correct" value. There is no such thing as the correct value of a sub-grid-model parameter.

---

[1] Parameterizations really come in two very different forms. The first are sub-grid models, which capture effects which would be captured by the main model if only the grid were small enough. The latter capture effects which would not be part of the model physics at any scale. An example of the latter might be plant growth rates given as a function of temperature, moisture, light, etc. It is somewhat unfortunate that these two kinds of modeling elements go under the same name. In this book, when I use the term, I am usually referring to the first kind of parameterization.

### d)  *Internal Variability*

As anyone who has ever experienced any weather knows, the climate system exhibits a fair amount of internal variability. It is hotter in Tampa today than it was yesterday, but not (one would assume) primarily because of a change in the season, or in the climate, or because of a change in a forcing, but because the system itself randomly fluctuates from day to day, week to week, and even year to year and beyond (see Chapter 5). But we should be careful about the sense in which this is a source of uncertainty about the climate. The fact that modelers are unable to predict the weather in Tampa for next February 23 does not, directly, mean that they have uncertainty about the climate, since knowledge of the climate does not entail knowledge of daily, weekly, or even annual weather fluctuation patterns. Knowledge of the climate only entails knowing the expected value (in the statistical sense) of any variable. Ignorance of the dynamics of internal variability does not, directly, diminish our ability to have this. You will read in various places that internal variability, or initial condition uncertainty, is a source of climate uncertainty. But such claims are sometimes misguided, and are sometimes phrased in such a way that obscures the above point.

Why is it sometimes correct to say, then, that internal variability can lead to climate uncertainty? Internal variability can contribute to uncertainty about the climate indirectly. This is because without doing a huge number of runs of a model, one for each possible initial condition of the model, and each for very long time periods, it is not trivial to determine what a model is predicting about climate, and what it is predicting about weather. For example, if I run one climate model for the next hundred years under some forcing condition, and I look at the average temperature over the last decade of the model, and I find some difference from today, is the difference attributable to a change in the climate, or merely a "random" (in the sense of internal variability) change in the weather between now and then? Enough systematic study of the model and under a variety of initial conditions can, to be sure, definitively settle this question; but this could easily be prohibitively expensive in every sense of the term. So the internal variability of the model can be a source of uncertainty.

This source of uncertainty is sometimes called "initial condition uncertainty" because it reflects uncertainty about whether a measured change (in a model) is due to the forcing or due to the choice of initial conditions. Importantly, as I have remarked elsewhere in the text (especially in Chapter 5), initial condition uncertainty is not uncertainty about (the actual, true) initial conditions. *Uncertainty about the true actual initial conditions can only lead to uncertainty about the weather, not the climate!* But internal variability of even a perfect climate model that was not systematically explored could lead the users of that model to have actual uncertainty about the climate.

### e) Real Data

Climate uncertainty due to uncertainty about real data, or observations, might refer to one of three things. First, it might mean uncertainty about what initial conditions to seed a climate model with. But as we have seen in our discussion of internal variability, the relevant uncertainty in climate projection is not about what the true initial conditions are. So uncertainty about the true initial conditions does not lead to climate uncertainty. Second, it might refer to the fact that climate models are compared against real data from historical records, (both from weather instruments and from so-called proxy data – see Chapter 2), and those data already have margins of uncertainty associated with them. A model that fits historical data perfectly might be less good than it appears if those historical data are themselves inaccurate. This second kind of real data uncertainty, therefore, only manifests itself in terms of uncertainty about model structure or parameter values – and so it is not an intrinsically separate source of uncertainty.

Models, however, are not the only source of knowledge in climate science and hence they are not the only source of uncertainty. There is a third way that uncertainty about real data leads to climate uncertainty. Paleoclimate data, for example, are a direct source of some of our best estimates of equilibrium climate sensitivity. And paleoclimate data are uncertain. This means that uncertainty about real data, which there sometimes is a great deal of, can lead to uncertainty about climate hypotheses directly.

### f) Boundary Values

Boundary values (such as the topography of the ocean floor, the degree of solar radiation, etc.) also play an important role in climate models and can implicate the model in varying degrees of uncertainty, since we might be uncertain about what the best way to represent these boundary values is, or because it might even be a significantly distorting idealization to even represent them *as* boundary values.

### g) Economic Scenario

It is often said that climate predictions involve uncertainties due to our lack of knowledge of future economic development, land use, population, etc. This is of course true, but since, strictly speaking, *climate projections* themselves are conditional on these factors, we should strictly speaking not include them as a source of uncertainty for climate projections. Climate projections, estimates of climate sensitivity, and estimates of presently observable climate change are either independent of, or conditional on, scenarios.

To summarize: it is likely that the overwhelming majority of the error in the projections made by our best climate models is due to three kinds of sources: model structure (understood as the combination of the original model equations, the discretization procedure, and the choice of parameterization schemes), choice of parameter values, and initial condition uncertainty (understood as the effect of potentially underexplored internal variability). Since knowledge of how to expect the climate to behave also comes from non-modeling sources (e.g. when we estimate climate sensitivity from paleoclimatalogical data), uncertainty in real data is also a direct source of uncertainty in general, and an indirect source of uncertainty in model projections (because the models are benchmarked against those real data). A fifth (probably smallest) source is uncertainty in boundary values.

## 7.3 How Uncertainty Is Quantified I: Ensembles and Sampling

Much of the effort in contemporary climate science to measure uncertainty from the first two sources (model structure and parameter values) is carried out by employing what one might call *sampling methods*. In practice these methods are extremely technically sophisticated. But this is mostly because of the high computational cost of each model run. Because of that high cost, modelers need to use sophisticated methods to reduce the number of model runs needed to explore the space of parameters and the space of models[2]. In principle, however, the overall methods are conceptually rather straightforward:

---

[2] One popular, (and fascinating!) technique for getting around the computational expense of running a model repeatedly to explore parameter space is to use "emulators": statistical simulations of climate simulations that are designed to approximate the output that a simulation model would have for a set of parameter values if we had the computational time to run it.

Take *parameter uncertainty*: consider the example of a simulation model with one parameter and several variables.[3] If one has a data set against which to benchmark the model, one could assign a weighted score to each value of the parameter based on how well it retrodicts values of the variables in the available data set. Based on this score, one could then assign a probability to each value of the parameter. Crudely speaking, what we are doing in an example like this is observing the frequency with which each value of the parameter is successful in replicating known data – how many of the variables does it get right? with how much accuracy? over what portion of the time history of the data set? – and then assigning this observed-frequency value to the probability of the parameter taking this value. The members of this family of methods for exploring parameter space are called "perturbed-physics ensemble methods."

The case of structural model uncertainty is similar. The most common method of estimating the degree of structural uncertainties in the predictions of climate models is a set of sampling methods called "multi-model ensemble methods." The core idea is to examine the degree of variation in the predictions of the existing set of climate models; the set of models that happens to be on the market, or those that are collected in an intercomparison project like the CMIP (see Box 2.2). By looking at the average prediction of the set of models and calculating their standard deviation, one can produce a probability distribution for every value that the models calculate.

Most multi-model ensemble methods and perturbed-physics ensemble methods have in common that they attempt to objectively (or perhaps more accurately: mechanically) calculate the degree of uncertainty from each source, respectively, by trying to systematically sample the range of possible results. If 80 percent of the results from a space of models and parameter values lie in some range, then the probability of the true result lying in that range is said to be 80 percent. Of course, they are *objective* only in the sense that they are independent of the degrees of belief of any particular expert, and that they are calculated mechanically.

Still, it is clear that what underwrites a large part of the motivation for using sampling methods like those described above is a desire to produce "objective" estimates of uncertainty, on some understanding of the term,

---

[3] Remember that a parameter for a model is an input that is fixed for all time, while a variable takes a value that varies with time. A variable for a model is thus both an input for the model (the value the variable takes at some initial time) and an output (the value the variable takes at all subsequent times). A parameter is simply an input.

and to avoid the stigma of subjective probability. The idea, presumably, is that by producing probability estimates that reflect observed frequencies of model outputs, one is measuring probability objectively. There are reasons to doubt, however, that these kinds of motivations and the methods that follow from them are conceptually coherent. I'll focus here on multi-model ensembles but perturbed-physics models suffer from some of the same problems when they are carried out in a similar manner.[4]

Signs of these problems are clearly visible in the results that have been produced. These signs have been particularly well noted by climate scientists Claudia Tebaldi and Reto Knutti (2007). They report that, in the first instance, many studies founded on the same basic principles produce radically different probability distributions. They show a comparison of four different attempts to quantify the degree of uncertainty associated with the predictions of climate models for a variety of scenarios, regions, and predictive tasks, and highlight the wide range of the various estimates. They summarize their conclusions in the following quotation:

> This study outlines the motivation for using multimodel ensembles and discusses various challenges in interpreting them. Among these challenges are that the number of models in these ensembles is usually small, their distribution in the model or parameter space is unclear, and that extreme behavior is often not sampled ...While the multimodel average appears to still be useful in some situations, these results show that more quantitative methods to evaluate model performance are critical to maximize the value of climate change projections from global models. (Tebaldi and Knutti 2007, p. 2053)

Indeed, there are four reasons to suspect that these mechanical methods are not a conceptually coherent set of methods for producing objective estimate of probability (in any sense of the term):

1) Ensemble methods either assume that all models are equally good, or they assume that the set of available methods can be relatively weighted.
2) Ensemble methods assume that, in some relevant respect, the set of available models represent something like a properly random sample of independent draws from the space of possible model structures.
3) Climate models have shared histories that are very hard to sort out.
4) Climate modelers have a herd mentality about success.

---

[4] Not all of the problems with multi-model ensembles also affect perturbed-physics ensembles, but the analog of problem 1, below, is especially acute.

Consider the following analogy. Suppose that you would like to know the length of a barn. You have one tape measure, and many carpenters. You decide that the best way to estimate the length of the barn is to send each carpenter out to measure the length, and you take the average. But there are four problems with this strategy. First, it assumes that each carpenter is equally good at measuring. But what if some of the carpenters have been drinking on the job? Perhaps you could weight the degree to which their measurements play a role in the average in inverse proportion to how much they have had to drink. But what if in addition to drinking, some have also been sniffing from the fuel tank. How do you weight these relative influences? Second, you are assuming that each carpenter's measurement is independently scattered around the real value. But why think this? What if there is a systematic error in their measurements? What if they are all using the same tape measure, and there is something wrong with it? Third (and relatedly) what if all the carpenters went to the same carpentry school, and they were all taught the same faulty method for what to do when the barn is longer than the tape measure? And fourth, what if each carpenter, before she records her value, looks at the running average of the previous measurements, and if hers deviates too much, she tweaks it to keep from getting a reputation as a poor measurer?

All of these sorts of problems play a significant role in making ensemble statistical methods in climate science conceptually troubled:

### 1) Ensemble Methods Either Assume That All Models Are Equally Good, or They Assume That the Set of Available Methods Can Be Relatively Weighted

If you are going to use an ensemble of climate models to produce a probability distribution, something in your background knowledge ought to support the claim that the models should be given equal weight in the ensemble. Failing that, you ought to have some principled way to weight them. But no such thing seems to exist. While there is widespread agreement among climate scientists that some models are better than others, quantifying this intuition seems to be particularly difficult. It is not hard to see why.

Gleckler, Taylor and Doutriaux (2008) display a beautiful figure (their figure 3) that shows color-coded metrics of predictive success for various models on various predictive tasks. It is fairly clear that while there are some unambiguous flops on the list, there is no unambiguous winner, nor a clear way to rank them. As Gleckler et al. make the point, no single metric of success is likely to be useful for all applications.

An analog of this problem exists for perturbed-physics ensembles as well. If we remind ourselves that parameterizations are such that there is no such thing as the true value of a parameter, then it is easy to see that any particular parameter value might also be good at one predictive task and less good at another.

### 2) Ensemble Methods Assume That, In Some Relevant Respect, the Set of Available Models Represent Something Like a Sample of Independent Draws from the Space of Possible Model Structures

This is surely the greatest problem with ensemble statistical methods. The average and standard deviation of a set of trials is only meaningful if those trials represent a random sample of independent draws from the relevant space – in this case the space of possible model structures. Many commentators have noted that this assumption is not met by the set of climate models on the market. In fact, I would argue, it is not exactly clear what this would even mean in this case. What, after all, *is* the space of possible model structures? And why would we want to sample randomly from this? After all, we want our models to be as physically realistic as possible, not random. Perhaps we are meant to assume, instead, that the existing models are randomly distributed around the ideal model, in some kind of normal distribution. This would be an analogy to measurement theory. But modeling isn't measurement, and there is absolutely no reason to think this assumption holds.

### 3) Climate Models Have Shared Histories That Are Very Hard to Sort Out

One obvious reason to doubt that the last assumption is valid is that all of the climate models on the market have a shared history. Some of them share code; scientists move from one lab to another and bring ideas with them; some parts of climate models (though not physically principled) are from a common toolbox of techniques, etc. But we do not have a systematic understanding of how much of climate model variation is systematic, how much of it is random, or indeed how much of it exists. So, it is not just the fact that most current statistical ensemble methods are naive with respect to these effects, but also that it is far from obvious that we have the background knowledge we would need to eliminate this "naivety" – to account for them statistically.

### 4) Climate Modelers Have a Herd Mentality About Success

This is a frequently noted feature of climate modeling. Most AOGCMs and ESMs are fairly tunable, and to the extent that no climate lab wants to be the oddball on the block, there is significant pressure to tune one's model to the crowd. This kind of phenomenon has historical precedent. In 1939 Walter A. Shewhart published a chart of the history of measurement of the speed of light (Shewhart and Deming 1939). His chart is a textbook case of herd mentality: despite the fact that researchers were nowhere near the correct value, using a variety of techniques, they all found approximately the same result.

## 7.4  How Uncertainty Is Quantified II: Expert Elicitation

If I am right that ensemble sampling methods for quantifying uncertainty are grounded and conceptualized out of a misguided desire to produce objective probabilities (whatever that might mean – since it can't mean objective *chances*), then an obvious alternative is an approach that embraces the fact that the probabilities in climate science are *personal*. Such an approach should *self-consciously* reflect the subjective degrees of belief of the relevant set of experts.

A now rather famous case of the IPCC moving in this direction was in its assessment of the probabilities regarding global mean surface temperature projections for 2100 in the AR4. Consider a projection contained in the technical summary for the AR4 for one particular emissions scenario – the details don't matter very much. The IPCC said that it was "likely" that, for the relevant scenario, global surface temperature would rise between 2.6 and 4.8°C relative to a twentieth-century baseline. Now, we know from the table in the previous chapter that this means that they considered the hypothesis of a rise between 2.6 and 4.8°C to be somewhere between 66 percent and 100 percent likely. What makes this interesting is that ensemble and sampling methods using the ensemble of models in the CMIP3 (the kinds of "mechanical" methods discussed above) had delivered an estimate of this probability that was much closer to 90 percent. Normally a 90 percent likelihood is what the IPCC would call "very likely."

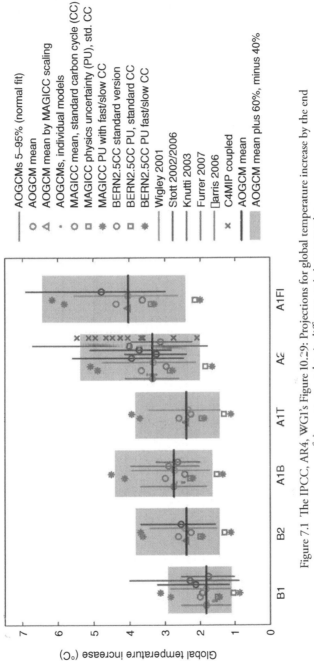

Figure 7.1 The IPCC, AR4, WG1's Figure 10.29: Projections for global temperature increase by the end of the century under six different emissions scenarios.

101

Wendy Parker provides a nice discussion of the relationship between the "mechanically produced" 90–100 percent probability and the figure arrived at by the IPCC:

> The discussion associated with the figure (figure 10.29 from the AR4) suggests that this uncertainty estimate was based on a number of relevant considerations, including the several ensemble studies whose results are also depicted in the figure, the experts' beliefs about the limitations of those studies (e.g. limitations in their treatment of carbon cycle uncertainties), as well as the experts' broader physical understanding of the climate system (see IPCC AR4, p. 810). The details of the process by which the specified ranges were determined to be likely, however, were not given.
>
> Unfortunately, such details often are omitted, making it difficult to evaluate the procedures used. In fact, at present, it is not entirely clear what the best-available methods for arriving at coarser estimates of uncertainty regarding future climate change are or could be. Nevertheless, whatever their details turn out to be, we can expect that they will prompt scientists to take into account all relevant data/results as well as any recognized limitations (i.e. biases, shortcomings, etc.) of the studies in which those data/results were produced. (Parker 2014, p. 29)

As Parker recounts it, two features of the procedure used to arrive at the estimate of probability are especially salient: 1) It is based on all the relevant sources of knowledge that are available: modeling results; known facts about the models and their shortcoming; inferences from observed data; expert knowledge about the quality of that data; etc. 2) It is based on the judgment of experts concerning what those sources tell us, and how they should be integrated with each other; not merely on a mechanical amalgamation of the sources.

## 7.5 Group Credences

In addition to the two Parker mentions, there are two other elements of the evidentiary landscape that she (canvassing the IPCC's discussion) does not, but that should surely be part of what a thorough debriefing of experts ought to elicit (continuing the numbering from above): 3) One could gather information on the degree of variation that exists in estimates of uncertainty – both among experts themselves as well as among the methods and data sets that the experts are consulting. 4) One could gather information on how resilient the experts expected their estimates to be in the face of what they would consider to be new data or modeling results that might be on the horizon.

The facts regarding 1) and 2) could be used to estimate both the degree of uncertainty regarding a hypothesis and how wide of a range of uncertainty to assign to it. Thus the facts regarding 1) and 2) led the experts who contributed to the AR4 to conclusions not only regarding the likelihood of global surface temperature rising between 2.6 and 4.8°C (under the relevant scenario) but also regarding how coarse-grained of an estimate to give (i.e. rather than locating the hypothesis as being 90 percent likely, they located it as being *at least* 66 percent likely.

Let us now return to a problem we raised in Chapter 6: philosophers usually take credences to be psychological characteristics of individual agents. The IPCC, however, is not an individual agent. What interpretation can we give to a report of credences from a group?

There are two different questions lurking here. The first is a question of how to understand what, for example, the IPCC actually did when it arrived at the above estimate in AR4. This is an empirical and historical question. It is one that we probably don't have adequate data to answer (at least: I don't have that data). And, it might very well be one for which, if we knew the historically correct answer, we would not find anything which we would find philosophically satisfactory.

It follows that there might be a better question. It should be better in the sense that we can actually answer it; and also better in that the answer we might give to it could be more philosophically satisfying. The better question is this: What *should* be the relationship between the individual expert opinions of the members of the relevant panel, and what the IPCC reports? One possible answer is that the answer ought to *reflect the consensus of the panel.* The panel should report a range of probabilities that is satisfactory to almost all, if not all, of the members of a panel. Group credences, in other words, should be considered to be the number, or range of numbers, that is the consensus of the group regarding what one's degree of belief in the hypothesis ought to be. We will come back to this point in the next chapter after we have discussed some further features of personal probabilities, especially the role of Bayes' rule.

## 7.6 Confidence

Anyone who has ever served on a committee knows that consensus can be reached by a variety of different paths. Sometimes it comes easily, with little discussion or haggling, and sometimes it involves making difficult compromises and accommodations. This brings us to the second-order qualifications that the IPCC attaches to some of the hypotheses it evaluates: those that express degrees of *confidence* in the first-order *probabilistic* claims.[5]

The most careful discussion by the IPCC that one can find of these matters is in a "Guidance Note for Lead Authors of the IPCC Fifth Assessment Report on Consistent Treatment of Uncertainties" which was released almost five years before the AR5. In that document, it was decided that

> The AR5 will rely on two metrics for communicating the degree of certainty in key findings:
> - Confidence in the validity of a finding, based on the type, amount, quality, and consistency of evidence (e.g., mechanistic understanding, theory, data, models, expert judgment) and the degree of agreement. Confidence is expressed qualitatively.
> - Quantified measures of uncertainty in a finding expressed probabilistically (based on statistical analysis of observations or model results, or expert judgment). (Mastrandrea et al. 2010, p. 2)

The AR4 had used slightly different language:

> A careful distinction between levels of confidence in scientific understanding and the likelihoods of specific results … allows authors to express high confidence that an event is extremely unlikely (e.g., rolling a dice twice and getting a six both times), as well as high confidence that an event is about as likely as not (e.g., a tossed coin coming up heads)." (AR4, 1.6)

There are two claims being made in this quotation. The first one is helpful, but the second is misleading. The useful point is that confidence and likelihood should be kept distinct, making it possible for there to be a confident declaration of a low likelihood, and vice versa. The second claim, however, is that the distinction between likelihood and confidence is really the distinction between objective chance and credence. What makes it possible to have a high confidence in a low probability is that one has a high *credence* in the objective fact that a *chance* is low. But this can't be the right way to think about confidence in the IPCC reports if objective chances play little to no role in climate science. Thus the guidance note for the AR5 is more useful, since it highlights, instead, that confidence ought to reflect "the type, amount, quality, and consistency of evidence (e.g., mechanistic understanding, theory, data, models, expert judgment) and the degree of agreement."

---

[5] This section benefited from a blog post by Wayne Myrvold and a discussion that ensued including myself, Kenny Easwaran, and Carl Hoefer (www.rotman.uwo.ca/two-dimensions-of-knowledge/)

Suppose, in other words, that a policy maker is consulting a group of experts in order to set her credence regarding a hypothesis. The group is going to report a probability, or a range of probabilities, to the policy maker based on a consensus they can reach. The policy maker might very well want to know:

- how many different sources of evidence were consulted by the experts in arriving at the assessments of probability.
- how univocal (or the contrary) those various sources were.

These are the two factors mentioned in the guidance note, but there really ought to be a third:

- the degree to which the reported consensus of the committee papered over internal disagreement or, to the contrary, reflected easy-to-come-by agreement.

All three of these factors are, in principle, independent. There could be many sources of evidence that disagree, or few sources that agree. There could be agreement or disagreement among the experts either way. Thus, in principle the policy maker might want to use all of this information as a guide to how she should act. She might use that added information, in conjunction with the reported probabilities, in a variety of ways. More conveniently, though, the policy maker might want the committee to summarize these three components of information into one single metric. One way to think about this kind of self-assessment of confidence is as the committee's assessment of the degree to which the answers to the above questions foretell a resiliency in their credences; as an assessment of how likely their consensus regarding appropriate credences is going to remain fixed in the light of future developments (be they in modeling, physical understanding, data acquisition, etc.).

One possible way to understand measures of confidence, therefore, might be as a kind of second-order probability – since, in effect, it would reflect the panel's estimate of the likelihood of their credence changing in the future. *A high confidence in a credence is a bit like a high degree of belief that that credence will be resilient in the face of future evidence* – assessed by looking at the variety of evidence supporting the credence, and the degree of agreement among those sources and among experts. But given the general murkiness of second-order probabilities in general, the lack of an obvious set of decision rules to apply to them, and the difficulties that would be involved in interpreting such probabilities in this specific case, I'm inclined to think that it is wise of the IPCC to refrain from using the expression "probability" for its second-order characterizations, and to limit itself to qualitative characterizations of confidence. (And it is best, I think, that they do so without an accompanying chart converting those qualitative expressions into numbers, as they do with the first-order likelihoods. Confidence is better conceptualized without being given a measure.)

## 7.7 Conclusion

The value-free ideal of science comes closest to being achieved when climate scientists attach probabilities to their hypotheses, and when they do not lose sight of the fact that the relevant probabilities are *credences* and not chances. Attending to the fact that they are credences involves eschewing attempts to rely on mechanical procedures for calculating probabilities, and moves instead to procedures that canvas the judgment of experts. That judgment should be conditioned on all the available sources of knowledge: modeling results; known facts about the models and their shortcoming; inferences from observed data; expert knowledge about the quality of that data; etc.; and on experts' best judgments concerning what those sources tell us, and how they should be integrated with each other. As we will see in the next two chapters, however, using probabilities to convey uncertainty can help us come closer to the value-free ideal, but it will not necessarily help us attain it. Sometimes probabilities can be a way for values to enter implicitly, and values can sometimes be buried under the probability assignments.

We also need to bear in mind that most philosophical discussions of credences invoke the mental states of individual agents, but both epistemic and political challenges faced by the IPCC and other groups of climate scientists make it such that it is best for them to report credences *as a group*, and often, to deliver ranges of probability rather than sharp values. As consumers of those reports, we should interpret those reports as reflecting the consensus of the relevant group of experts, and producers of the reports should do their best to deliver estimates of probability that are in line with that expectation.

Reporting a consensus range of probabilities does not, however, exhaust the relevant information that groups of experts can conveniently summarize for policy makers and other consumers of expertise. They can also offer assessments of the confidence that have in those probabilistic appraisals of hypotheses that summarize the variety of sources of evidence supporting the hypotheses, the degree of agreement in those sources, and the degree of agreement among the experts. Consumers ought to be able to interpret those degrees of confidence along the lines that would be reflected in taking them to be second-order probabilities: assessments of the likelihood that the experts' first-order credences will be revised in the light of novel evidence. But given the complexities of second-order probabilities, and the degree to which that model of "confidence" *papers over its underlying multi-dimensionality* (variety of evidence; agreement in sources of evidence; agreement among experts), the IPCC is wise not to use the word "probability" in connection with its levels of confidence and to report them only qualitatively, and without a conversion chart from qualitative expressions to ranges of quantities.

In the next two chapters, we turn to two interrelated questions. First, how can consumers of these kinds of reports use them to make decisions (regarding policy and other matters)? Second, how well does all of this help climate science meet the value-free ideal of science?

## Suggestions for Further Reading

Mastrandrea, Michael et al. 2010. "Guidance Note for Lead Authors of the IPCC Fifth Assessment Report on Consistent Treatment of Uncertainties." A new document supporting the IPCC, AR5, on the treatment of uncertainties, which also includes a treatment of the subject of this chapter: confidence.

Parker, Wendy. 2010a. "Predicting Weather and Climate: Uncertainty, Ensembles and Probability," *Studies in History and Philosophy of Science Part B: Studies in History and Philosophy of Modern Physics*. A very nice philosophical overview of ensemble methods in weather and climate prediction with a discussion of how they should be interpreted.

Tebaldi, Claudia and Reto Knutti. 2007. "The Use of the Multi-Model Ensemble in Probabilistic Climate Projections," *Philosophical Transactions of the Royal Society of London A: Mathematical, Physical and Engineering Sciences*. A detailed scientific discussion of ensemble methods for quantifying uncertainties in climate science.

# 8

# *Decision*

## 8.1 Statistical Inference: Two Paradigms

In some of the discussions that follow, it will be helpful to have a basic understanding of two contrasting paradigms of statistical inference – the Bayesian paradigm and the classical paradigm.

A simple example can illustrate the differences rather well: suppose that we work at a factory that produces belt buckles. The proprietor of the factory assigns us the task of evaluating hypotheses regarding the belt buckles produced in the factory and the extent to which they are defective. Imagine further that what we have at our disposal is the following sort of device. The device sits at the end of the production line and as each belt buckle passes through it, one of three indicator lights is illuminated: one that says "Defective" (D), one that says "Sound" (S), and one that says "Misaligned" (M).

We also have at our disposal a small collection of belt buckles that we know to be sound and a small collection that we know to be defective, and we notice the following behavior of the detector. Regardless of what kind of buckle we place in the detector, 10 percent of the time the detector reads "M" (presumably indicating that it is unable to test the buckle because it did not enter the detector box in the correct manner). For cases in which the "M" light does *not* go off, we observe the following (stable) frequencies. When we have placed a known-to-be-sound buckle in the device, it reads "S" 98 percent of the time and "D" the other 2 percent. And when we have placed a known-to-be-defective buckle in the device it reads "D" 98 percent of the time and "S" the other 2 percent.

Our first task is to construct a test to determine whether a particular belt buckle should be tossed into the recycling bin when it comes off the line. One obvious choice would be to recycle any buckle that produced either a "D" or an "M."[1] Let us define three terms.

---

[1] We could also only throw out buckles that produce a "D." The reader can easily reconstruct the reasoning below for such a case.

1) Type I error (sometimes called "a false negative,"): throwing away a perfectly good belt buckle.
2) Type II error (sometimes called "a false positive"): keeping a defective buckle.
3) The "power" of our test: the probability that we will throw away a particular buckle if that buckle is defective. (This is just 1 minus the probability of type II error.)

We can easily calculate the probability of type I error, pr(I); type II error, pr(II); and the power.

Pr(I) = .1 + .9 × .02 = .118; because that is the probability of getting "D" or "M" with a sound buckle.

Pr(II) = .9 × .02 = .018; because that is the probability of getting "S" with a defective buckle.

Power = .1 + .9 × .98 = .982; because that is the probability of getting a "D" or "M" when a buckle is defective.

We might also want to construct a procedure for testing the hypothesis that our factory only produces sound buckles. Here is one: we treat the hypothesis that the factory only produces sound buckles as the null hypothesis. We test 2,000 buckles. If more than 260 of the buckles we test set off either the "D" or the "M" indicator, we reject the null hypothesis.

We would now want to define type I error as rejecting the null hypothesis even though it is true. (Our hypothesis is now about the factory in general, not about a particular buckle.) The probability of that, in the above, is .046. (The calculation is a bit more complicated, but only because there are more cases to keep track of.) If this particular test (where we require 261 or more "D"s or "M"s) told us to reject the null hypothesis, we could say that our test result was significant to less than .05 (which is usually considered, in classical statistics, as the criterion of good evidence that the null hypothesis is false.) It is significant at that level, because it has only a 4.6 percent chance of being wrong.

Conversely, we would now define type II error as failing to reject the null hypothesis even though it was false. The probability of this would depend on what the actual rate of defective buckles coming off the production line was. If the actual rate of defective buckles was 1 percent, the probability of type II error would be 0.309. So we would call the power of this test (which in this case would be its power to detect a failure rate of at least 1 percent) .691.

These three kinds of statements of probability are what we might call "classical probability statements." They tell us the probabilities of getting various measurement outcomes conditional on the true state of the object or system being measured. They tell us, for example, the probability of getting a "D" or "M" if a buckle is sound. Most importantly, *none of them are statements concerning the probabilities of the true states of the system.* We have not, in any of these statements, said anything, for example, about the probability that a particular buckle is sound given any set of measurements. Thus, in the classical paradigm of statistical inference, we rest content with knowing facts of the following kind (which we might usefully call "reverse probabilities"): if this particular buckle were defective, how likely or unlikely would it be for me to get the measurement outcome that I got? In other words, the classical statistician makes statements of "significance." If she observes a measurement outcome that is very unlikely (the usual threshold for unlikeliness is less than 5 percent likely) when some "null hypothesis" is true, then she reports that the null hypothesis can be rejected with such and such significance, where significance is the (hopefully low) probability of getting the measured outcome given the null hypothesis. In our test above, for example, if the null hypothesis is that a buckle is sound, getting a reading of "D" or "M" only allows us to reject the null hypothesis at the relatively poor significance of .118. (I say, "relatively poor" because the value is considerably higher than the traditional 5 percent, or .05). Importantly, the classical statistician never tells us the probabilities of the hypotheses that interest us.

The Bayesian finds all of this to be terribly unsatisfactory. The Bayesian wants to be able to make probabilistic claims about the true state of the world given particular measurement outcomes, not just the reverse. (All of the statements above were "reverse probabilities" – they were about the probability of a measurement outcome given a particular state of the world.) For example, the Bayesian would like to say what she thinks is the probability of a belt buckle being defective given that the device registered a "D." How does she do this, and why does the classical statistician object to this?

The answer, unsurprisingly, has to do with an equation from which Bayesians get their name, called "Bayes' rule," which is helpful for "inverting" conditional probabilities.

A conditional probability, first of all, is the probability of some event (or proposition) conditional on the occurrence (or truth) of some other event (or proposition). In our example above, for instance, we might be interested in the probability of getting "S" conditional on a buckle being sound.

We write this as "pr("S"|Sound)" and we define it mathematically as

$$pr(\text{"S"}|\text{Sound}) =_{\text{def}} pr(\text{"S" \& Sound})/pr(\text{Sound}).$$

Just from this definition and from the axioms of probability, we can derive the Bayes rule (instantiated below with our example):

$$pr\left(\text{Sound}|"S"\right) = \frac{pr("S"|Sound) \times pr(Sound)}{pr("S")}$$

where the right-hand side can be rewritten by breaking up *pr*("S") into as many mutually exclusive and exhaustive states as we like. For us it will be useful to use Sound and ~Sound. So we get:

$$pr\left(\text{Sound}|"S"\right) = \frac{pr("S"|Sound) \times pr(Sound)}{pr("S"|Sound) \times pr(Sound) + pr("S"|\sim Sound) \times pr(\sim Sound)}$$

Notice that in the last equation, we have everything we need to calculate $pr\left(\text{Sound}|"S"\right)$ except for the value of *pr*(Sound). Bayesians call this value the "prior probability of Sound" because it is the probability that they would give to Sound *prior to finding out* that the instrument read "S." Calculating the "posterior probability" $pr\left(\text{Sound}|"S"\right)$ from the prior probability of Sound and the "likelihood" of "S" given Sound, $pr("S"|Sound)$, using Bayes' rule, is what Bayesians call "conditionalization." The idea that conditionalization is the primary principle regarding how to update your beliefs in the light evidence (or sometimes, even, that it is the only rationally required form of belief modification) is one of the core tenants of Bayesianism.

Suppose, for example, that our Bayesian belt buckle inspector would like to know the probability that some belt buckle was Sound, having had the detector read "S" when it passed through. If she assumes that, prior to measurement, the buckle was equally likely to have been Sound as Unsound (.5 each), she can make the calculation.

It is just:

$$pr\left(\text{Sound}|"S"\right) = \frac{.98 \times .5}{.98 \times .5 + .02 \times .5} = .98$$

But assuming that the buckle was equally likely to have been Sound as Unsound was a very substantive assumption. She might, instead, have had much greater confidence in the manufacturing procedure, and thought that the buckle was 99 percent likely to be sound, prior to measurement. If so, her new calculation is

$$pr\left(\text{Sound}|"S"\right) = \frac{.98 \times .99}{.98 \times .99 + .02 \times .01} = .9998$$

and if she has very low confidence in the manufacturing procedure, such that she thinks the buckle was only 1 percent likely to be sound, prior to measurement, then she gets:

$$pr\left(\text{Sound}|"S"\right) = \frac{.98 \times .01}{.98 \times .01 + .02 \times .99} = .3333$$

This is a huge difference. While the classical statistician will be forced to either accept or reject the null hypothesis that the buckle was sound upon seeing the device read "S," and report a confidence level of .118, the Bayesian can give a *probability for the hypothesis*, and it might be anywhere (in the examples we saw) from 33.33 percent to 99.98 percent.

Oversimplifying just a little bit, the difference between the classical statistician and the Bayesian comes down to this: the Bayesian finds the classical statistician's report to be inadequately informative and inadequately sensitive to what she might already know. And the classical statistician is skeptical of the Bayesian's pronouncements about her prior probabilities. Where, the classical statistician will ask, do these come from?

## 8.2  Probability and Inference

The latter point, more often than not, comes down to the attitudes each paradigm has regarding what we called "credences" in Chapter 6. Notice that, more often than not, probabilities of measurement outcomes given states of the world can be regarded as objective chances. That's because physical measurements are, in general, chance setups. I can repeatedly put sound belt buckles into my measurement device, and I am likely to get stable frequencies of "S"s and "D"s. So I can regard the probability of getting "S" on a sound buckle as an objective chance.

This is generally not the case regarding prior probabilities of hypotheses. In our toy example, to be sure, the production of belt buckles might also be a chance setup. And so we could in this particular case regard the prior probability of a buckle being sound as an objective chance. But *in general*, this is not the case with scientific hypotheses. Consider a hypothesis like the one that vaccine xyz prevents disease D. If my null hypothesis

$H_0$ is that the vaccine has no effect on the rate of disease contraction, then it's plausible to think that the probability of finding the disease in a particular person who has had the vaccine, conditional on $H_0$, is an objective chance: vaccinating a person and exposing them to real-world conditions is a chance setup that produces stable frequencies of outcomes. So pr(contract disease|$H_0$) could be an objective chance. The classical statistician wants to hang her hat on this. But it is hard to see what it might mean to say that the objective chance of the hypothesis being true is pr(H).

The Bayesian has no problem with this, since she will simply assert that her prior probability that xyz prevents D is just her credence – the personal degree of belief she had in the hypothesis prior to the evidence coming in. But the classical statistician balks at this. This is why people who pay a lot of attention to credences are sometimes called "Bayesians," and why some Bayesians insists that all probabilities are credences. But this is something of a confusion of two orthogonal points. Classical statisticians can also believe in credences and Bayesians can also believe in chances. For our purposes, it suffices to note two things.

1) The classical statistician will never report probabilities of hypothesis conditional on the evidence. She will simply accept them or reject them based on the evidence and on pre-established procedures and pre-established confidence levels. This is *usually* because she is skeptical of prior probabilities.
2) The Bayesian will happily report probabilities of hypotheses (conditional on the evidence) but can only do so at the cost (in most cases) of freely admitting that her probabilities are personal credences.

In response to worries (coming from the classical camp) about the apparent "subjectivity" of 2), Bayesians are often keen to point out two things. The first is that the priors assigned by different agents to a hypothesis may vary considerably, but (assuming the agents agree on what counts as evidence and how such evidence bears on the priors assigned to a hypothesis) it typically (that is to say in the absence of pathological examples) takes very little evidential input to get the posterior probabilities to converge. The second point is to employ some intellectual judo on the classicists. The Bayesian will point out that the classicist ignores almost all prior information and fails to incorporate what we already know. Worse still, the Bayesian will point out that the classicist relies just as heavily on subjectively determined choices as the Bayesian, but sweeps them under the carpet rather than owning up to them the way the Bayesian does. Those choices include the null hypothesis, the space of outcomes, the p-level to

use for significance, and the stopping rule. All of those are as much subjective choices as are priors for the Bayesian.

Let's return now, briefly, to the question we raised in the last chapter
about probabilities for a hypothesis assigned by a group. The idea of group
probabilities puts some pressure on the usual "purist's" conception of
credences, according to which credences are degrees of belief. Agents, on
the purist's view, do not estimate probabilities, they just have them. The
role of Bayes' rule, in such a conception, is to represent how a perfect agent
would respond to certain pieces of information – perhaps as a way of determining the extent to which an actual agent meets the ideal. But there are of
course situations in which a scientist might realize that the degrees of belief
that they happen to have do not reflect the best available evidence. Or
they might just not have degrees of belief in the purist's sense: they might
have no particular disposition to bet on climate hypotheses. Even worse, if
probabilities are just things that you have, it is difficult to see how a group
can deliberate together about what the correct ones are. Following Cohen
(1995) and Steel (2015), therefore, I think we should think of personal
probabilities as things that scientists accept. On this view, the careful scientist will use whatever methods of reasoning are at her disposal to correct
the degrees of belief that she happens to have, and in the case of the model-
based science we are discussing, she will attempt to estimate what the best
probabilities for her to use are, based on her best models. And she will use
Bayes' rule as a way of calculating probabilities, rather than as a constraint
on rationality. On this second view, it makes more sense to think of groups
of scientists collectively deliberating about what personal probabilities to
adopt regarding a hypothesis.

## 8.3  Risk and Uncertainty – Three Approaches

Unfortunately, in the case of climate science, things might not be so rosy
for the Bayesian. As we have seen (in Chapter 7), climate scientists are
often unable, or at least collectively unwilling, to assign precise and determinate probabilities to the relevant states of the world. What is the probability that equilibrium climate sensitivity (ECS) is in the range 1.5°C to
4.5°C? The IPCC is only willing to say that this probability is 66 percent *or
higher*. As to the probability that ECS is greater than 6.0°C, the IPCC says
10 percent or less, but only says this with "medium confidence."

When precise probabilities do not exist, there are three possible courses
of action available to the decision maker: 1) delay action until better information is available, 2) demand precise probabilities from her experts, or

3) use alternative (non-standard, i.e. non-EU) decision rules (Bradley and Steele 2015). Each of these has obvious disadvantages.

1) Delaying action is, in most cases, not actually possible. In most cases, delaying action is likely to have an outcome of its own, often one that is not unlike one of the policy choices you were considering in the first place. If the question, for example, is whether or not to take a putative cancer-fighting drug, delaying action is nearly equivalent to deciding not to take the drug.

2) What about demanding more precise probabilities from your experts? This approach is strongly defended by John Broome.

> The lack of firm probabilities is not a reason to give up expected value theory. You might despair and adopt some other way of coping with uncertainty ... That would be a mistake. Stick with expected value theory, since it is very well founded, and do your best with probabilities and values. (Broome 2012, p. 129) (quoted in Bradley and Steele 2015, p. 810)

One might worry, however, that demanding more precise probabilities than experts are comfortable with providing, is a recipe for asking them to make decisions for you on matters of value or utility. This is the opposite of what they are supposed to do on the division of labor that decision theory is supposed to provide ... This is especially true if you believe, as many do, that certain "paradoxes" of decision making (the "Allais paradox" and the "Ellsberg paradox") reveal flaws in the constraints on rational action that are embodied in expected utility theory (EU). Here is why:

Let's start with the Allais Paradox. Both so-called paradoxes are in fact experimentally revealed preferences in human decision making that are inconsistent with EU. (This is not by itself, of course, the makings of a genuine paradox in the philosopher's sense of the term.) The observed behavior arises when comparing participants' choices in two different gambling scenarios that the participants are offered. The Allais experiment involves the following two gambles being offered to the participants in the experiment (Table 8.1).

Many studies, some with these exact payoffs, others with real monetary payoffs that are equivalent to these, and even others with examples such as health outcomes, support the claim that most people prefer to choose gamble 1A over 1B, but that they also prefer gamble 2B over 2A. Look carefully at the four columns. The only difference between column 1A and column 2A is that 89 percent of the probability of getting the outcome of $1 million has been replaced with the outcome of getting nothing. But the very same is true of the difference between columns 1B and 2B. Column

Table 8.1 *The Allais paradox experiments*

| Gamble 1 | | | | Gamble 2 | | | |
|---|---|---|---|---|---|---|---|
| Choice 1A | | Choice 1B | | Choice 2A | | Choice 2B | |
| Winnings | Chance | Winnings | Chance | Winnings | Chance | Winnings | Chance |
| $1 million | 100% | $1 million | 89% | Nothing | 89% | Nothing | 90% |
| | | Nothing | 1% | $1 million | 11% | | |
| | | $5 million | 10% | | | $5 million | 10% |

Table 8.2 *The Allais paradox experiments rewritten*

| Experiment 1 | | | | Experiment 2 | | | |
|---|---|---|---|---|---|---|---|
| Choice 1A | | Choice 1B | | Choice 2A | | Choice 2B | |
| Winnings | Chance | Winnings | Chance | Winnings | Chance | Winnings | Chance |
| $1 million | 89% | $1 million | 89% | Nothing | 89% | Nothing | 89% |
| $1 million | 11% | Nothing | 1% | $1 million | 11% | Nothing | 1% |
| | | $5 million | 10% | | | $5 million | 10% |

1B has an 89 percent probability of the outcome $1 million, and that has been replaced in 2B by adding 89 percent to the chance of getting nothing.

This can be seen most clearly by rewriting the chart as below, where the payoffs are exactly the same, but some of them have been broken into two different squares just to create more visual symmetry between 1 and 2.

If you ignore the replacements of the dark-grey squares with the light-grey squares, which are identical in both columns (1A differs from 2A exactly as 1B differs from 2B), then the gambles offer the gamblers exactly the same choices. Therefore, a single agent making choice 1A and 2B cannot be consistent with expected utility theory no matter what utilities that agent assigns to each marginal dollar. What some take this to show is that many agents will sacrifice expected utility in order to avoid the great sense of disappointment that would come from having chosen 1B and getting nothing, having known that $1 million was a sure thing in choice 1A. It also seems to show that in addition to assigning their own utilities to specific outcomes, agents can have different preferences regarding risk.

Table 8.3 *The Ellsberg urn paradox*

| Gamble 1A | Gamble 1B |
|---|---|
| You receive $100 if you draw a red ball | You receive $100 if you draw a black ball |

| Gamble 2A | Gamble 2B |
|---|---|
| You receive $100 if you draw a red or a yellow ball | You receive $100 if you draw a black or yellow ball |

The Ellsberg "paradox" bears more direct relevance to the suggestion (under consideration here) that we demand of experts that they deliver more precise probabilities. The kind of "paradox" in question is again simply an empirical finding that seems to conflict with the demands of EU. In this case, the experiment involves an urn with 90 balls in it. 30 of the balls are red, and the other 60 are a mix of black or yellow. Participants are not told the total number of black balls nor the total number of yellow balls, only that the total of the two is 60. They are then offered the following two gambles – each from the same kind of urn (described above).

The situation involves what is sometimes known as "Knightian Uncertainty," named after the economist Frank Knight, who wrote:

> Uncertainty must be taken in a sense radically distinct from the familiar notion of Risk, from which it has never been properly separated ... The essential fact is that 'risk' means in some cases a quantity susceptible of measurement, while at other times it is something distinctly not of this character; and there are far-reaching and crucial differences in the bearings of the phenomena depending on which of the two is really present and operating ... It will appear that a measurable uncertainty, or 'risk' proper, as we shall use the term, is so far different from an unmeasurable one that it is not in effect an uncertainty at all. (Knight 2012, p. 19)[2]

The reason that these two gambles exhibit Knightian uncertainty is fairly clear: the gambler does not have a good way of settling on a probability of getting a black ball or the probability of getting a yellow ball. She

---

[2] Knight distinguished his kind of uncertainty from "risk," but this vocabulary is not universal ("uncertainty" often just means probabilities that are neither 0 nor 1), so one tends to use the term "Knightian uncertainty" to avoid confusion. Another possible confusion: people sometimes use the phrase "Knightian uncertainty" to refer to any situation in which outcomes are not associated with an objective chance, rather than restricting to situations in which agents cannot arrive at a *determinate* subjective credence. I use it here in the stronger sense. The Ellsberg setup can of course be seen as instantiating either one.

has a probability for getting a red ball (1/3), and she has a probability for getting-a-yellow-ball-or-black-ball (2/3), but not the probability for each of those separately. What experiments reveal is that most people strictly prefer gamble 1A to 1B, but also strictly prefer 2B to 2A. This violates expected utility theory as well. Why? Since all of the payoffs are identical, the preference for 1A indicates, according to the theory, that the gambler assigns a higher probability to getting a red ball than to getting a black ball. But if that is true, then whatever probability she assigns to getting a yellow ball, adding it to both sides of the bet shouldn't alter her choice. But it does. She prefers 2B to 2A – presumably because now it is the right-hand choice that has a determinate probability. What many people believe this shows is that such a gambler (who is typical) has *"ambiguity aversion."* That is, she has a preference, to some degree, for betting outcomes regarding which there is no Knightian uncertainty. In this case, the gambler seems to value the red balls strictly more than black balls and strictly more than the yellow balls, even though she does not value the yellow-and-black balls combined less than twice as much as the red balls. One way of thinking about this is that the gambler is putting more decision weight (by the lights of EU) on red than she assigns probability to it, precisely because she has no "ambiguity" about that probability. She also seems to be putting more decision weight on black-or-yellow than on red-or-yellow, without any conviction that this outcome is more probable. She takes decision weight away from black and yellow, when they occur by themselves, because she has ambiguity about what those respective ambiguities ought to be. And she takes decision weight away from red-or-yellow for the same reason.

How does this relate to situations in which experts are reluctant, either individually or as a group, to assign precise probabilities to a hypothesis? Suppose, for example, that you know nothing of the behavior of balls and urns, and you consult an expert to give you the probabilities of getting a red ball, a yellow ball, and a black ball, respectively. And suppose the expert follows the dictum that you should always bite the bullet and deliver precise probability estimates, on the grounds that only those can be plugged into the decision rules of expected utility theory. And suppose, finally, that you have a degree of ambiguity aversion that the expert does not have access to. In such a situation, the expert cannot give you a precise probability estimate for each of the three outcomes that is not in some way prejudicial regarding your tastes. If the expert declares that there is a 33.3 percent probability of red, and apportions the remaining 66.7 percent of the probability mass to black and yellow, she will be implicitly saddling you with a commitment to no ambiguity aversion. But she cannot shift decision mass to the red ball outcome without prejudging what degree of ambiguity aversion you ought to have.

3) Finally, we should consider the possibility of employing alternative decision rules to those that seek to maximize expected utility. One clear disadvantage of this route is the absence of consensus regarding what such rules should be. In general, the only thing that is almost universally agreed on regarding what constraints such an alternative set of decision rules should abide by is this: it is widely agreed that if a particular choice is "dominated" according to EU, then it had better not come out as the correct choice according to a proposed set of alternative decision rules. For our purposes, what it means for a choice to be dominated according to EU is this: suppose you are uncertain as to what probabilities to assign to various possible outcomes, but, like the IPCC, you are willing to assign a range of probabilities to each of them. A choice is dominated according to EU if, for any particular set of probabilities of outcomes that fall within that range, EU tells you that the choice in question is a bad one. Suppose, in other words, that you have an urn with 30 red balls, and some unknown number of yellow balls; the latter could be anywhere from 0 to 70. You therefore take the probability of a red ball to be 0.3 and the probability of a yellow ball to be somewhere in the range [0, 0.7]. Suppose, moreover, that you are offered a bet that offers you A if a red ball is selected, and nothing otherwise, and one that gives you B if a yellow ball is selected, and nothing otherwise. And suppose finally that you value A three times as much as B. In such a case, choosing the second bet would be dominated by the choice of the first bet according to EU because the first bet would come out as the better one no matter where in your allowed range the probability of a yellow ball were taken to be. So what is widely agreed on is that no alternative decision theory should be taken seriously if it delivered the verdict, in such a situation, that you chose the second bet.

Unfortunately, this constraint is not very strong. In effect, it only requires that everyone agree on a course of action when EU says that the degree of Knightian uncertainty is not decision-relevant. Alternative decision rules become more controversial to the extent that they often try to encapsulate the idea that a decision maker ought to be *cautious* to some degree or other, and much of what's controversial about these rules can be seen as disagreements about what degree of caution is warranted. One very cautious rule, for example, is the maximin rule, which says that one should always choose the course of action whose worst possible outcome is better

than all the other choices' worst possible outcomes (the "maximum min-
imal outcome"), regardless of what the probabilities (or range of possible
probabilities) is associated with those outcomes. But this rule could be
unnecessarily cautious – and it pays no attention to coarse-grained probability distributions at all.[3] Many alternative decision rules designed for
decisions under Knightian uncertainty, moreover, fail to account for information that can be encapsulated in confidence assessments of the kind we
discussed in Chapter 8.

Arguably, climate science and climate policy are in sore need of a
decision-making framework for decisions under Knightian uncertainty
that is appropriately sensitive to all of these issues: probabilities that
reflect the degrees of belief of groups of experts; coarse-grained probability estimates; those that come with confidence assessments; the need
for a decision framework whose embedded degree of caution is appropriate to reasoning about global policy and global environmental risks;
etc. Given that the development of such a framework could depend on a
good understanding of probability, social epistemology, the epistemology
of complex models, decision theory, topics in global justice, and political
philosophy, this could be a rich area for philosophers from a number of
sub-disciplines to work on.

## 8.4  Probability and Decision

As we noted in Chapter 6, the advantages of scientific hypothesis evaluation
coming in the form of probabilities, rather than in the form of acceptances
and rejections, are significant. As any consumer of scientific knowledge
is aware, being told that a null hypothesis has failed to be rejected can be
frustratingly unhelpful information. Suppose, for example, that you have a
debilitating disease, and a new drug is available that carries with it a known
and significant set of unpleasant side effects. Suppose, furthermore, that
the drug has been tested on a relatively small sample of people, and the
null hypothesis, that the drug provides no benefit against the disease, failed
to be rejected. What should you do? It is hard to know, because this gives
you no guide at all to the probability that the drug provides a benefit. But
if you had a good estimate of the probability that the drug provided some

---

[3] Of course, one might reasonably argue that given the severity of the threat of various forms of climate change, that one could never be *too* cautious in the climate case. But that, of course, is already
  to make a value judgment.

Table 8.4 *Associating a "utility" (the Oi) with each state/action pair*

| State (drug effective?)<br>Action (take drug?) | Yes | No |
|---|---|---|
| Yes | $O_1$ | $O_2$ |
| No | $O_3$ | $O_4$ |

benefit, you could employ the standard theory of decision making under risk by employing expected utility maximization (hereafter EU) to decide whether to use the drug.

EU associates actions undertaken in possible real underlying states of the world with quantified consequences. We can think of these quantified consequences as measures of the utility the decision maker assigns to the complete outcome she associates with performing that action in that underlying state of the world. So for example, in the case of choosing whether or not to take a drug, EU would assign a utility to four different action/outcome pairs.

If the consumer can quantify the degree of utility she associates with each of these four outcomes, and she knows the probability that the drug is effective, pr(eff), then she can decide how it is rational to act. She should take the drug just in case

$$(pr(eff) \times O_1) + ((1 - pr(eff)) \times O_2) > (pr(eff) \times O_3) + ((1 - pr(eff)) \times O_4)$$

The left-hand side of the inequality is the "expected utility" of taking the drug, and the right-hand side is the expected utility of not taking it, so she should take the drug just in case the former is greater than the latter. But if the consumer is only given the results of classical statistical tests, and only takes the drug if the null hypothesis that the drug is ineffective has been rejected, then her own utilities that she associates with each of the four outcomes play no role in her decision making. It is therefore fair to say that in choosing a null hypothesis, a space of outcomes, a significance level and a stopping rule, the classical statistician has swept not only her priors, but also her implicit assignments of the utilities associated with each of the four outcomes, under the rug. The classical statistician has also taken the ability to make use of her own utilities out of the hands of the consumer

## 8.5 Cost Benefit Analysis

Some readers may be wondering what the last two sections have to do with climate *science*[4]. Indeed, as I will articulate in more detail in the next chapter, part of the point of probabilism in science and of the utility maximization framework of decision theory is that it facilitates a division of labor into the epistemic and the normative. Science can deliver probabilities, and consumers of that scientific knowledge, presumably stakeholders in the climate system, can choose how to act in the light of those probabilities using any set of preferences they like. This is good, presumably, because we tend to think questions about what we take to be in our best interests are questions about our values, and those are not scientific questions.

Consider a proposition that would seem to fall squarely on the side of being a decision on how to act, rather than a matter of scientific fact: the proposition that we ought to hold "the increase in the global average temperature to well below 2°C above pre-industrial levels and to pursue efforts to limit the temperature increase to 1.5°C above pre-industrial levels, recognizing that this would significantly reduce the risks and impacts of climate change." This is a proposition that was agreed on at the *21st Conference of the Parties of the United Nations Framework Convention on Climate Change in Paris* in 2015.

Many people would take it as relatively obvious that this is not, and indeed cannot be, a purely scientific conclusion. After all, for one thing at least, you might think that a mass extinction event that would wipe out the human species among many others would be a good thing. And it seems like it would be foolish to think that there is a scientific argument against that proposition. Even setting aside cartoon examples like this, and even if we assume some fairly broad agreement about what kind of future for the planet is desirable, it is unlikely that there is a clear scientific argument that 1.5°C and 2.0°C represent the precisely correct points at which to draw the relevant lines.

There are, to be sure, scientific facts that have a strong bearing on whether or not to accept the Paris recommendations. It's a scientific claim (and inter alia, one that the best science supports) that 2.0°C of warming represents a point at which many climate tipping points are likely to appear. And it's a scientific claim (again, a well-supported one) that 2.0°C of warming would take human beings outside the range of global temperatures that we have ever experienced in our 200,000 or so years of existence. But as

---

[4] Much of this section follows Frisch (2013, 2017)

Knutti et al make the point, "no scientific assessment ever defended or recommended a particular target." Such a conclusion will always reflect a combination of scientific findings as well as political agreements about what risks are worth taking, and what benefits are too costly to give up.

If all of this is right, then there is a sense in which discussions of how we should act in the light of what we know about the climate do not belong, strictly speaking, in a book on the philosophy of climate *science*. It would all seem to be, rather, a matter of climate *ethics*. But there is another group of voices who speak out about this topic, voices coming particularly from the world of economics, who believe that how we should act in response to the threat of climate change is a scientifically discoverable fact. They argue, moreover, that a certain kind of model, what is sometimes called an "optimization integrated assessment model" or IAM, can tell us how we should act – that is, what, if any, mitigation measures we should take to hold back climate change. Since a fairly influential group of people are making this claim, and are in fact making use of what purport to be scientific models in order to arrive at their conclusions, some discussion of their work is warranted in a book that is on climate science and not climate ethics.

Proponents of this approach believe that expected utility theory and cost-benefit calculations can provide a means of precisely calculating what the ideal climate mitigation strategy is. The basic idea is that the ideal climate mitigation strategy is the one that maximizes intergenerational expected utility. It proposes, in other words, that we balance the cost of mitigation strategies to ourselves against their benefits to future generations that they would accrue as the result of preventing damages that would otherwise result from greenhouse gases that would be emitted were it not for those mitigation efforts.

How can a model calculate the intergenerational expected utility of a particular mitigation strategy (and thereby be able to select the strategy that maximizes this)? In principle it is very simple. For each possible strategy, we need to know the economic cost of executing the strategy, and we need to know the future benefit that we would accrue from such a strategy in terms of the benefits of avoiding temperature increases. The former is more straightforward. The latter requires that we have, for each mitigation strategy, a probability distribution over different possible temperature outcomes, as well as an estimate of the loss of utility associated with each such outcome.

In practice, the way this is done is to couple an extremely simple climate model to an economic model so as to represent the impacts of climate change on human welfare, the impact of changes in economic activity on greenhouse gas (GHG) emissions, and the effect of mitigation measures on economic growth.[5] The coupling of the two models is duplex, that is: the output of the economic model affects an input to the climate model by having the economic activity predicted by the former result in GHG emissions, which affect the latter; the output of the climate model affects an input to the economic model in the form of a "damage function." The damage function relates the degree of warming predicted by the climate model to a loss of economic growth in the economic model.

What this means, again in practice, is that these models need to assume at least four things. The first concerns the impact of emissions on climate change. The second concerns the impact of climate change on economic activity (the damage function). The third is a welfare model, which is a function that sums up all of the things that contribute to the overall welfare of global societies as a whole. And the fourth is a function that relates the importance of future losses (or gains) against present costs (or benefits.) The fourth function is usually specified in terms of a "discount rate," which "tells you" the present value of future losses. If, for example, your "discount rate" is 3 percent, then any future gain or loss in the economy that occurs, for example, in 50 years, is worth $1.03^{50}$ times as much as that today, or 22 percent. In other words, at a discount rate of 3 percent, I should only be expected to spend $22 today in order to save future generations $100 in 50 years.[6]

All four of these components pose significant challenges to the idea that cost-benefit analysis is a "scientific" alternative to the combination of scientific and political agreements that are embedded in the Paris goals. We can see this by taking a look at each one.

### 1) The Discount Rate

The discount rate answers the question: how should we weight some particular value that might be enjoyed by our future descendant relative to how much we would weight a similar value to ourselves? This might seem like a minor problem, but in fact the recommended course of action that we get from IAMs is extremely sensitive to the discount rate that we use.

---

[5] Examples of these include William Nordhaus's DICE model (Nordhaus 2008), the PAGE model, which was used in the Stern Review (Stern 2007), and Richard Tol's FUND model (Tol 2002a, Tol 2002b).

[6] The idea of a discount rate comes from the realm of finance, where one is interested in calculating the attractiveness of an investment opportunity by comparing the amount invested today with the expected return at some future date.

The range of values used in various studies is large. The Stern review used a discount rate of 1.5 percent while others used values as high as 5.5 percent. This is one of the largest factors responsible for whether a study based on an IAM recommends modest or stringent mitigation policies. My own view is that the correct discount rate to use would be to set the value to the per capita rate of economic growth in the model. If future generations are going to be 10 percent richer than us per capita, then we should value our own present wealth 10 percent more than we value theirs[7]. And vice versa. This is a simple consequence of adopting a kind of egalitarian reasoning. Again, however, since this is a book on climate science and not climate ethics, I am not going to take up much space arguing for that here. I simply want to point out that there is no scientific, as opposed to ethical, argument that one can make for this claim.

## 2) *The Impact of Emissions*

The climate impact model component of most IAMs is extremely simple. Often what they use is either a simple value for ECS[8], or a probability distribution over a range of possible values. Now, of course, the question of what the correct value of ECS is is a scientific question, and not an ethical one. However, as anyone who has read this far into the book should understand, it is a scientific question about which there is a great deal of uncertainty. That uncertainty, moreover, is Knightian uncertainty. And as I argued in section 8.4 above, knowing how we should make decisions under Knightian uncertainty requires us to know more than simply utilities and the relevantly coarse-grained probability distribution over the possible outcomes. It also requires us to know how the relevant agents value risk and ambiguity. In other words, even if you know what utility I assign to different possible outcomes, and even if you have a coarse-grained probability distribution over those possible outcomes, you cannot determine, scientifically, how I ought to act without knowing how cautious I am in the face of risk and uncertainty (as the Allais and Ellsberg "paradoxes" reveal.)

Consider, for example, the fact (which we will discuss in detail in sections 12.3 and 12.4) that the IPCC considers is "very unlikely" (with

---

[7] In other words, we should set the discount rate to the annualized value of that 10 percent rate of growth, whatever it happens to be.

[8] Equilibrium Climate Sensitivity (ECS) is discussed in several places in the book, including section 8.4 of this chapter, but also 3.4, 4.3, extensively in Chapter 6 and Chapters 11 and 12.

medium confidence) that ECS is less than 6.0°C. This means that they assign as much as 10 percent of their probability weight to values higher than 6.0°C and consider it possible that, in the light of new evidence, they could add even more probability weight to that. This represents a fairly large degree of both risk and Knightian uncertainty regarding a very dangerous part of the ECS curve. Given the rational possibility of risk and ambiguity aversion, not everyone's values will be consistent when it comes to weighting their utilities regarding this part of the curve directly in proportion to the probability mass. It is a perfectly rational preference to overweight this area of the curve out of caution. So picking a particular probability distribution over values of ECS and then classical decision theory should be applied to it is not a scientifically warrantable conclusion. It simply assumes without argument that the relevant agents are neutral with regard to risk and ambiguity.

### 3) The Damage Function

The damage function is the part of the model that correlates warming with loss of economic growth. The DICE model proposed by (Nordhaus 2008), for example, represents the coupling from the climate to the economy with a single mathematical function of the form $C(T)=1/[1+(T/a)b]$. DICE then goes further and assumes that $b=2$, meaning that it assumes the function is quadratic, i.e. a polynomial of order 2. But the correct form that such an equation should take is highly empirically underdetermined. The reason is simple: we only have data relevant to this question over a very small part of the curve. There is no paleo-proxy data regarding the economy, for example. And there are no physical principles that provide guidance about what order the equation ought to be. Given those facts, there will be third order polynomials and fourth order polynomials, etc., that fit the available data just as well as Nordhaus's. But these undetermined alternative functions differ with Nordhaus's most where it matters most: at the most dangerous end of the curve. As Frisch points out, given an 8°C increase in global temperatures, the DICE and PAGE models predict about a 15 percent damage (reduction) to global gross domestic product (GDP). The FUND model predicts damages of roughly 6 percent GDP. One might reasonably wonder whether these predictions are overly optimistic, given that higher order polynomials that fit existing data equally well would predict much higher levels of damage.[9]

---

[9] Frisch also points out that more careful and detailed examinations of finer grained data in other studies (Burke et al. 2015) deliver much higher levels. of damage (up to an order of magnitude higher

## 4) The Welfare Model

The welfare model component of an IAM is supposed to represent the total overall welfare of everyone on the planet at a particular time. It is meant to be all-inclusive and represent the satisfaction of any preference that anyone values at any time. Obviously, many such preference satisfactions do not have a market price. Nordhaus (of the DICE model) is well aware of this and writes: "Economic welfare should include everything that is of value to people, even if those things are not included in the market place." (Nordhaus 2008, p. 13). In reality, however, IAMs almost all (including DICE) use GDP as a proxy for global welfare.

There are arguably two problems with this, which should be kept distinct. The first is the obvious problem that not all goods that we value are plausibly included in GDP, nor are they capable of being replaced by economically produced goods. If climate change were to kill off all the coral reefs in the world, and snorkeling at a coral reef is something I enjoy more than all the iPhones in the world, then production of more iPhones, no matter how many, is never going to satisfy my preferences as much.

Of course, defenders of IAMs have a response to this. They can argue that we cannot avoid comparing the value of economic and non-economic goods when we make policy decisions, and so, in practice, we have to act as if these goods can be compared mathematically (Frisch 2013). But even if we accept this argument, there is still a problem. While it may very well be true that, in policy making, we have to make these determinations, it doesn't follow that any particular choice of how to do this is scientific.

Perhaps more importantly, there is a second problem with using GDP as a measure of global welfare that is independent of the incommensurability of economic and non-economic goods. Even if we assume that two goods are commensurable, it does not follow that they are intersubstitutable as the world changes. Suppose, in the example above, you force me to put an economic value on my snorkeling access to coral reefs. Suppose I agree that one trip to the Great Barrier Reef is worth 25 iPhones. It is still inevitable that as more reefs are destroyed, and reefs become more scarce, their value compared to iPhones will go up. How much their value goes up will depend on our preference structures, not on any scientific facts.

---

for end-of-the-century prediction than DICE, for example.) My own point here is not to argue for the correctness of either of these predictions, but only to show that they are highly empirically underdetermined – much more so, than, for example, ECS projections.

## 8.6 Conclusion

None of the above are meant to be devastating criticisms of the very idea of IAMs. IAMs might very well be useful tools for thinking through policy decisions related to climate mitigation. But two points need to be kept in mind if they are to be used responsibly: 1) They need to be reasonable in assessing the relevant uncertainties regarding basic elements like ECS and the damage function, and 2) They cannot be presented as more scientific alternatives to politically deliberated goals like the Paris goals. Many of the inputs to IAMs (the discount rate, the degree of risk and ambiguity aversion, the value we set on non-economic goods, and the rate at which we treat them as going up as they become more scarce) are not matters of epistemic expertise. They are questions that can only be answered by querying the relevant stakeholders (many of whom, in this case, have yet to be born).

At the same time, however, we should be somewhat careful regarding this last point: that IAMs are not a scientific alternative to political deliberation. One might conclude from all of these remarks that in order for something to be considered scientific, it must be free of considerations

---

**Box 8.1 The Paris Agreement**

The Paris Agreement is an agreement regarding greenhouse gas emissions that begins to take force in the year 2020. The agreement was negotiated by representatives of 195 countries at the 21st Conference of the Parties of the United Nations Framework Convention on Climate Change (UNFCCC) in Paris. It was adopted by consensus on December 12, 2015 within the framework of the UNFCCC.

The aim of the convention is described in Article 2, "enhancing the implementation" of the UNFCCC through:

(a) Holding the increase in the global average temperature to well below 2 °C above pre-industrial levels and to pursue efforts to limit the temperature increase to 1.5 °C above pre-industrial levels, recognizing that this would significantly reduce the risks and impacts of climate change;
(b) Increasing the ability to adapt to the adverse impacts of climate change and foster climate resilience and low greenhouse gas emissions development, in a manner that does not threaten food production;
(c) Making finance flows consistent with a pathway towards low greenhouse gas emissions and climate-resilient development. ("The Paris Agreement" 2017)

of normative value. But we should carefully distinguish cases where the inputs to a model are explicitly normative (as they are in the case of an IAM), and cases where modeling choices are dependent on normative values in a much more subtle way. We turn to this second kind of case in the next chapter.

## Suggestions for Further Reading

Bovens, Luc and Stephan Hartmann. 2003. *Bayesian Epistemology*. An extremely useful reference on Bayesian methods in the sciences, as well as their application to solving problems in philosophy of science and epistemology.

Peterson, Martin. 2009. *An Introduction to Decision Theory*. An excellent philosophical introduction to decision theory – including good discussions of the interpretation of probability and Bayesian inference.

Bradley, Richard and Katie Steele. 2015. "Making Climate Decisions," *Philosophy Compass*. An outstanding discussion of decision theory in the context of climate science and climate policy making.

Broome, John. 2012. *Climate Matters: Ethics in a Warming World*. A lucidly written and pioneering introduction to the field of climate ethics.

Nordhaus, William D. 2015. *The Climate Casino: Risk, Uncertainty, and Economics for a Warming World*. Especially parts 3 and 4. A set of recommended policy responses to global climate change by Nordhaus derived from his model-based approach to cost benefit analysis of climate mitigation policy. Readers should be warned that many of Nordhaus's recommendations are considered overly optimistic by well-informed experts – for many of the reasons discussed in this chapter.

Frisch, Mathias. 2013. "Modeling Climate Policies: A Critical Look at Integrated Assessment Models," *Philosophy & Technology*. A comprehensive critique of model-based cost benefit analyses of climate mitigation strategies. Many of my arguments in section 8.5 follow Frisch.

# 9

# *Values*

## 9.1 Introduction

To what extent can and should scientific research reflect our social and ethical values?[1] It is uncontroversial that, to some degree, it does, and it must. When we set constraints on experimentation, for example, or when we decide which scientifically informed projects to pursue and which projects to ignore, such decisions uncontroversially reflect social and ethical values. If we decide not to test cosmetics on animals, it can only be because of considerations of ethics – for example, that we might wish to avoid animal cruelty more than we wish to have safer or cheaper or better performing cosmetics. And that can only be a reflection of our social and ethical values. And if a pharmaceutical company decides to allocate more resources to the development of cures for erectile dysfunction than for cures for tropical diseases, such a decision can only be criticized on the basis of social and ethical values – that it is ethically bad to make people suffer from malaria in exchange for a benefit we place a lower value on. It cannot be criticized as bad epistemology.

Importantly, decisions such as these are in a sense *external* to scientific research[2]. They do not, in any case, involve the *appraisal of hypotheses, theories, models, or predictions*. While they clearly involve value judgments, they do not intermingle with expertise in the same way that hypothesis appraisal does. The philosophically controversial question about values in science is about the degree to which values are involved (or better put: the degree to which they are necessarily involved, or inevitably involved, and

---

[1] This chapter borrows liberally from J. Biddle and Winsberg (2009), Winsberg (2012) and Parker and Winsberg (2017). Conversations with Biddle, Parker, and also Torsten Wilholt have played a large role in shaping my views on the role of values in science. Conversations and work with Johannes Lenhard (Lenhard and Winsberg 2010) have also played a role in informing my views on scientific modeling that are relevant to this chapter.

[2] Or at least so it is usually claimed. See more on this point in section 9.4.

perhaps most importantly: uncorrectibly involved) in the appraisal of hypotheses or in reaching other conclusions that are internal to science; and that necessarily also involves scientific expertise. The goal of this chapter is to see how the role that values play in climate science can advance the traditional debate about values in science

To begin to do that, we need to distinguish between the role that values do and should play inside scientific research from the ones that are external to it. But that's not all. We also need to distinguish between different kinds of values. What the traditional debate primarily concerns is the role of *social or ethical* values. These sorts of values, I take it, are the estimations of any agent or group of agents regarding what is important and valuable – in the typical social and ethical senses – and what is to be avoided, and to what degree. What value does one assign to economic growth compared to the degree to which we would like to avoid various environmental risks? In the language of decision theory, by social values we mean the various marginal utilities one assigns to events and outcomes (see Chapter 8). In the present context, the point of the words "social" and "ethical" is primarily to flag the difference between these values and what Ernan McMullin once called "epistemic values": simplicity, fruitfulness, empirical adequacy, etc. (McMullin 1982) According to McMullin's line, these are things that we value in a scientific hypothesis[3], but only because we take them to be truth-conducive. Hypotheses that we take to exhibit these values are more likely to be true. But I do not want to beg any questions about whether or not values that are paradigmatically ethical or social can or cannot, or should or should not, play important epistemic roles. So, I prefer not to use that vocabulary – the vocabulary of "epistemic" vs. "non-epistemic" values.[4] I talk instead about social and ethical values when I am referring to things that are valued for paradigmatically social or ethical reasons.[5]

This is a question of some importance. We would like to believe that only experts should have a say in what we ought to believe about the natural world. But we also think that it is *not* experts, or at least not experts *qua* experts, who should get to say what is important to us, or what is valuable, or has utility, etc. Such a division of labor, however, is only possible

---

[3] And for which you and I might assign different degrees of value. (I might value simplicity in a hypothesis more than you, and you might value fruitfulness more.)

[4] Another common piece of language is to call social or ethical values "contextual values." This is especially true in the influential work of Helen Longino.

[5] I do not carefully distinguish, in this chapter, between the social and the ethical And when I variously use the expressions "social values," "ethical values," or "social and ethical values," these differences in language should not be read as flagging important philosophical differences.

to the extent that the appraisal of scientific hypotheses, and other matters that require scientific expertise, can be carried out in a manner that is free of the influence of social and ethical values. It is also an important question because we do not want scientists to be engaged in "wishful thinking." We do not want a scientific enterprise in which our experts positively evaluate a hypothesis because, on their system of values, it would be desirable for that hypothesis to be true. So it is important to get a grip on how realistic the value-free ideal is – the ideal that scientists should appraise hypotheses in a manner that is independent of what social, ethical, and political values they hold.

Here's the plan for this chapter. In section 9.2 I will recount a now-famous debate in philosophy of science and statistics from the 1950s between Richard Rudner and Richard Jeffrey regarding what is now known as the "argument from inductive risk." Rudner suggested that the argument showed that values played an ineliminable role in science, but many observers at the time took Jeffrey to have shown that Bayesianism afforded a defeating response to Rudner's arguments. In section 9.4, I will look at some of the replies to Jeffrey that philosophers of science have mounted in the last decade or so. In 9.5 I will try to show how a close attention to climate science and scientific modeling can add to the list of replies available to those who would defend the claim that values play an ineliminable role in science. In 9.6 and 9.7, I will pursue some of the consequence of this for how the IPCC ought to report uncertainties.

## 9.2 Risk

Philosophers of science have mounted a variety of arguments to the effect that the epistemic matter of appraising scientific claims of various kinds cannot be kept free of social and ethical values.[6] Here, we will be concerned with one such line of argument and some variations on it. It is an argument that is closely connected to our discussion of uncertainty quantification in climate science (see Chapters 6 and 7) – and it goes back to the mid-twentieth-century work of a statistician, C. West Churchman (Churchman 1948, 1956), and a philosopher of science, Richard Rudner (Rudner 1953). This line of argument is now frequently referred to as

---

[6] In addition to Churchman and Rudner, mentioned below, see also Douglas (2009, 2000, 2005), Machamer and Douglas (1999), Longino (1990, 1996, 2002), Kourany (2003a, 2003b), Wilholt (2009, 2013), Elliott (2011), Elliott and McKaughan (2014), Elliott and Resnik (2014), Steel (2015, 2010).

the argument from *Inductive Risk*. It was articulated by Rudner in the following schematic form:

1) The scientist qua scientist accepts or rejects hypotheses.
2) No scientific hypothesis is ever completely (with 100 percent certainty) verified.
3) The decision to either accept or reject a hypothesis depends upon whether the evidence is sufficiently strong.
4) Whether the evidence is *sufficiently* strong is "a function of the *importance*, in a typically ethical sense, of making a mistake in accepting or rejecting the hypothesis."
5) Therefore, the scientist qua scientist makes value judgments.

Rudner's oft-repeated example was this: how sure do you have to be about a hypothesis if it says: 1) a toxic ingredient of a drug is not present in lethal quantity, vs. 2) a certain lot of machine-stamped belt buckles is not defective? As Rudner put it, "How sure we need to be before we accept a hypothesis will depend upon how serious a mistake would be" (Rudner 1953, p. 2). We can easily translate Rudner's lesson into an example from climate science. Should we accept or reject a hypothesis, for example, that, given future emissions trends, some regional climate outcome will occur? Should we accept the hypothesis, let's say, that a particular glacial lake dam will burst in the next 50 years? Suppose that if we accept the hypothesis, it will behoove us to replace the natural moraine that holds back the lake with a concrete dam. In such a case, whether we want to build the dam will depend not only on our degree of evidence for the hypothesis but also on how we would measure the severity of the consequences of building the dam, and having the glacier not melt, vs. not building the dam, and having the glacier melt. Thus, Rudner would have us conclude: as long as the evidence is not 100 percent conclusive, then we cannot justifiably accept or reject the hypothesis without making reference to our social and ethical values regarding floods and dam building.

The best-known reply to Rudner's argument came from the philosopher, logician, and decision theorist, Richard Jeffrey (1956). Jeffrey argued that the first premise of Rudner's argument, that it is the proper role of the scientist qua scientist to accept and reject hypotheses, is false. Indeed, he took Rudner's argument to be a reductio on the premise. Their proper role, he urged, is to assign probabilities to hypotheses with respect to the currently available evidence. Others, for example policy makers, can attach values or utilities to various possible outcomes or states of affairs, and in conjunction with the probabilities provided by scientists, decide how to act. (Using the decision-theoretic apparatus we discussed in Chapter 8.)

In providing this response to Rudner, Jeffrey was making clear a point that we made in Chapter 6: that an important purpose of probabilistic forecasts is to separate practice from theory and the normative from the epistemic. It is *part of the point* of probabilistic forecasting to see to it that social values can be relegated entirely to the domain of policy, and cordoned off from the domain of scientific expertise. If the scientist accepts or rejects a hypothesis, then Rudner has shown that normative considerations cannot be excluded from that decision process. In contrast, if scientists don't have to bring any normative considerations to bear when they assign probabilities to a hypothesis, and they refrain from accepting and rejecting hypotheses, then the normative considerations can be cordoned off.

It seems likely, however, that Jeffrey did not anticipate the difficulties that modern complex-model-based science would have with the assignment of probabilities to hypotheses with respect to the available evidence. There are many differences between the kinds of examples that Rudner and Jeffrey had in mind and the kinds of situations faced by, for example, climate scientists. For one, Rudner and Jeffrey discuss cases in which we are interested in the probability of the truth or falsity of a single hypothesis, but climate scientists generally are faced with having to assign probability distributions over a wide space of possible outcomes. Another significant difference between the classic kind of inductive reasoning Jeffrey had in mind (in which the probabilities scientists are meant to offer are their subjective degrees of belief based on the available evidence) and the contemporary situation in climate science, is the extent to which epistemic agency in climate science is distributed across a wide range of scientists and tools. And finally, I don't think that Jeffrey anticipated a kind of science in which off-the-shelf models, entities that do not have psychological states like degrees of belief, would be playing some of the roles usually played by experts. In any event, I believe climate science, and in fact any kind of science that relies intensively on the kinds of complex modeling efforts that are employed in climate science, involves added complexities vis-à-vis the debate on the role of values in science, and the goal of this chapter will be to explore those.

After the exchange between Rudner and Jeffrey in the 1950s, interest in inductive risk arguments in philosophy of science mostly waned, until they were brought back into the limelight by Heather Douglas (2000). In addition to reinvigorating the discussion, Douglas added a new volley: she noted a flaw in Jeffrey's response to Rudner. She remarked that scientists often have to make methodological choices that do not lie on a continuum. A simple example can illustrate her point. Suppose I am investigating the hypothesis that substance X causes disease D in rats. I give an experimental group of rats a large dose of X and then perform biopsies to determine what percentage of them has disease D. But how do I perform the biopsy? Suppose that there are two staining techniques I could use; one of them is more sensitive and the other is more specific. That is, one produces more false positives and the other more false negatives. Which one should I choose? Douglas points out that which one I choose will depend on my inductive risk profile. To the extent that I weigh more heavily the consequences of saying that the hypothesis is false if it is in fact true, I will choose the stain with more false positives. And vice versa. But that, she points out, depends on my social and ethical values. Social and ethical values, she concludes, play an inevitable role in science.

Inevitability, of course, is always relative to some fixed set of background conditions. The set of background conditions Douglas is assuming is one where something roughly like classical statistical methods are being used (see Chapter 8). If I have some predetermined level of confidence, alpha, say .05, and some predetermined stopping rule, then the choice of which staining method I use will raise and lower, respectively, the likelihood that the hypothesis will be accepted. Indeed, many of the points we made in section 7.3, regarding the conceptual difficulties confronting ensemble and sampling techniques for estimating climate uncertainties, parallel Douglas's points about methodological choices. If you mechanically aggregate the results you get from a set of staining methods (or climate models), the staining methods (or climate models) you choose will bake into the cake whether you are going to be more likely to err in one direction or another.

In Chapter 7 we suggested that eliciting expert judgment was a less conceptually fraught way to go about estimating climate uncertainties. This, in fact, is what any good Bayesian will always say. There is a parallel (more general) point to be raised in response to Douglas. What if, instead of employing a predetermined level of confidence, etc., all toxicologists were good Bayesians of the kind that Jeffrey almost surely had in mind? Why could they not use their expert judgment, having chosen whatever staining method they like, to factor in the specificity and sensitivity of the method when they use the evidence they acquire to update their degrees of belief about the hypothesis? In principle, surely they could. By factoring the specificity and sensitivity of the method into their degrees of belief, they should be able to factor out the influence of the social/ethical values that otherwise would have been present when a staining method was chosen. And if they could do this, social and ethical values, at least the kind that normally plays a role in the balance of inductive risks, would not *necessarily* play a role in their assessments of the probabilities.[7] Let us call this the Bayesian response to Rudnerian and Douglasian arguments from inductive risk (BRAIR). I call it the Bayesian response in part to highlight the fact that Bayesians *have always been keen* to point out that classical methods of inference, with their predetermined confidence levels and stopping rules, have always had a problem with sweeping matters of utility or values under the rug.

In what follows, I want to look at a variety of lines of response that arise in reply to BRAIR. In section 9.3, I will look at replies that have been offered by others in the literature, and then in section 9.4 I will look at a line of response that is more specific to climate science, or at least to sciences that involve complex modeling of roughly the kind we find in climate science.

## 9.3  Replies to the BRAIR

### *Second-Order Inductive Risk*

The Bayesian response is subject to various counter-replies. A standard one is that probability assignments themselves – and more generally estimates of uncertainty – are also subject to inductive risk considerations (Rudner 1953, Douglas 2009). If a scientist concludes that H's probability is P, this just pushes the problem one layer up, since we now have to accept a hypothesis H' about the probability of H (that it is P). So, the argument goes, one has not escaped the need to decide whether evidence is strong enough to warrant accepting a hypothesis. It's just that now, the hypothesis in need of acceptance is H' rather than H. Such an argument would not be uncontroversial. First, it assumes that an agent can be uncertain about what probability to assign to H. Second, it assumes that this sort of uncertainty is subject to being overcome by evidence of a certain strength. This seems to assume that there are second-order probabilities – i.e. probabilities about probabilistic hypotheses.

---

[7] Whether they would do so in fact is not what is at issue here. Surely that would depend on features of their psychology and of the institutional structures they inhabit, about which we would have to have a great deal more empirical evidence before we could decide. What is at stake here is whether their social and ethical values would *necessarily* play a role in properly conducted science.

There is of course a deeply psychologistic story one can tell about credences on which all of that is incoherent. On that story, credences are just things that people have; they are nothing more than the dispositions people have toward various betting possibilities. On such a story, it makes little sense to be uncertain about what your own credences are. But recall that in Chapter 8 we suggested that such a story about credences is a poor fit for understanding the probabilities that the IPCC and others assign to climate hypotheses. We agreed to follow Cohen (1995) and Steel (2015) in thinking of the personal probabilities of climate scientists as things that scientists do indeed accept and reject. On this view, the careful scientist will use whatever methods of reasoning are at her disposal to correct the degrees of belief that she *just happens to have*. If this is the right way to think about credences in climate science, then Steel is right that, for real-world cases, social and ethical values can have a role to play deep in the heart of Bayesian analyses. Real-world epistemic agents, he suggests, often simply do not have precise degrees of belief to serve as the priors and likelihoods needed to run a Bayesian analysis; instead, agents decide how to represent these probabilities – using distributions that are uniform, binomial, Poisson, normal, etc. All of this suggests that Rudner was right in his "second-order" reply to Bayesians like Jeffrey.

## *The External vs. the Internal Workings of Science*

Another reply to the BRAIR is to question something that we have been taking for granted. It is fairly standard in the science and values literature to concede that decisions about what evidence to collect counts as "external" to science. But should it always? Nobody doubts, after all, that we have collected more data about the earth's atmosphere than the moon's,[8] in part because of what we deem socially important. But in many cases, deciding what data to collect, and when to stop collecting more, can have a strong influence on what probabilities we will assign to hypotheses.[9] Consider an obvious case: suppose it is 1925 and most people are assigning a low probability to the hypothesis that smoking causes cancer. Should we spend a certain amount of money to study the question of whether the hypothesis is true? The answer to this depends on obviously social and ethical values. (What else would the money be spent on? How much do we value those other things compared to knowing the truth about smoking?)

---

[8] Yes, the moon does have an atmosphere!
[9] This point was driven home to me by Torsten Wilholt in personal correspondence.

At the same time, the answer to the question of whether we should spend the money will have an obvious impact on the degree of belief we assign to the hypothesis. Of course the 1925 scientists cannot predict whether the research will raise or lower the probability of the hypothesis, but they can predict that it will very likely push that probability in one direction or another. This blurs the line between the epistemic and the normative in a way that most commenters seem to believe won't happen if the values only play a role that is "external" to science.

## 9.4  A Model-Specific Response to BRAIR: Prediction Preference

One thing that should be relatively clear from the preceding section is that I think much of science is fated to fall short of the value-free ideal. Whether it is because scientists are using classical statistical methods that bake values into the cake, because they have to make methodological choices that are not on a continuum, because they make choices about what probabilities to accept, or because they make decisions about how much data to collect, social and ethical values *will* very often play an uncorrected role in science. They will play an uncorrectable role in shaping what hypotheses we accept if we choose to play the game of accepting and rejecting; or they will play an uncorrectable role in shaping what probabilities we assign to hypotheses if we chose to be probabilists. So I should be clear that *I do not regard climate science to be special in its failure to meet the value-free ideal.* Nevertheless, I do think that climate science – and other sciences in which large, complex simulation models that are built by a large network of scientists play a central role – have features that pose a special challenge to meeting the value-free ideal. As we will see, one thing that makes the challenge different from all the ones I have described above is that the values in question are not necessarily "inductive risk" values – they do not necessarily reflect the preferences one has for making one kind of error compared to another. Rather, they will often reflect "prediction preference": the preferences we have for making one kind of prediction accurately compared to another.

### *The Argument*

The argument that model-intensive science cannot be free from the influence of the value we place on successfully making one kind of prediction over another begins with a truism about model evaluation we already discussed in Chapter 3: models are usually accepted on the basis of being considered adequate-for-purpose, not on the basis of being considered true, or even on the basis of anything like considerations of "overall fit."

A single scientific model of a phenomenon or system, moreover, cannot and will not serve all possible purposes. For example, for a model to provide very accurate predictions, it may need to be quite complex, but that same complexity will make it a poor model for teaching, or for illustrating a simple causal relationship. When we build a scientific model, we usually do it with a set of purposes in mind, to some of which we attach a higher degree of *importance*. For example, a weather model might be constructed with the primary aim of forecasting where hurricanes will make landfall, because those forecasts will inform a set of decisions that need to be made about whom to evacuate, where to send emergency crews, how to apportion resources, etc., and with the secondary aim of gaining better understanding of the basic weather patterns that drive hurricane migration. Turning back to climate models, different climate models might be developed on the basis of the prioritization of different aims. In climate model development, achieving different kinds of model skill is often a trade-off: a model can be structured so as to better resolve Madden-Julian oscillations in the tropical atmosphere, but at the expense of not doing as well at forecasting tropical cyclone formation[10] – or vice versa.

In other words, the purposes and priorities that a model builder has in mind will influence decisions in model construction and evaluation (see also Elliot and McKaughan 2014; Intemann 2015). When it comes to model construction, for instance, purposes and priorities shape not just which entities and processes are represented in a model but also *how* they are represented, including which simplifications, idealizations and other distortions are more or less tolerable. This is particularly true if a model is parameterized, as almost all climate models are, because model parameter values are chosen by optimizing the model's performance. How you define performance – in terms of which prediction tasks are most important to you – will therefore influence what parameter values you chose.

The same point is true, mutatis mutandis, about model evaluation. When you evaluate a climate model's "skill," you are always doing this relative to some particular prediction task or, at least, to a specific weighted combination of several such tasks. In practice, of course, model construction

---

[10] This was an actual example that was given to me by a climate scientist when I asked for an example of model-prediction trade-off. He reported that his group had chosen tropical cyclone formation because it was a "hot topic" for the upcoming CMIP.

and model evaluation are not neatly separable. Much informal evaluation occurs during the process of model construction, and what is learned via formal evaluation activities typically feeds back into the model construction process. For example, if results for a variable that significantly influences Madden-Julian oscillations fits relatively poorly with observations, then adjusting the model to achieve better results for that variable may become a high priority if correctly predicting those oscillations was a priority. Such adjustment might involve increasing the fidelity with which a particular process is represented or including in the model a representation of a process that before had been omitted entirely.

The next point is particularly crucial. In certain kinds of modeling-intensive sciences, with climate science being a particularly acute case, models are developed in a layered way. Climate modelers never build their models from scratch. As we noted earlier, most climate models are descendants of weather models. And in general, the latest climate models are usually made by modifying existing models to meet some particular task. When that is done, it is impossible to undo the effects of all the previous model-building choices. In such cases, purposes and priorities from earlier times can continue to exert an influence. Suppose process A in a model that is being constructed was represented in a particular way at an earlier time in the model's development because of priorities at that time. That A was represented in that way might well have shaped how some other processes were subsequently represented in the model; perhaps process F was represented in a way that helps to compensate for some errors introduced by A's representation. A decade later in the model's development, priorities may have shifted somewhat, but modelers probably will not bother to change the way A is represented, unless there is a good reason to do so. This is because they know that, were A to be given a different – even higher fidelity – representation now, the model might perform less well; such a change might upset the "balance of approximations" that currently allows the model to successfully simulate some important features of the target system or phenomenon (see also e.g. Mauritsen et al. 2012, p. 14). Such interdependencies among errors are often fully expected, even though exactly what the dependencies are is often unclear, and so earlier layers of model development are left intact (see also Lenhard and Winsberg 2010).

Finally, it is important to note that it will be very difficult if not impossible for climate scientists to completely and adequately keep track of the exact impact of these past modeling choices on their present models. This is especially true because of three features of climate models:

## 1) Size and Complexity

Climate models are enormous and complex. Take one of the state-of-the-art American models, the NOAA's GFDL CM2.X (see Box 4.4). The computational model itself contains over a million lines of code. There are over a thousand different parameter options. It is said to involve modules that are "constantly changing," and involve hundreds of initialization files that contain "incomplete documentation." The CM2.X is said to contain novel component modules written by over one hundred different people. Just loading the input data into a simulation run takes over two hours. Using over a hundred processors running in parallel, it takes weeks to produce one model run out to the year 2100, and months to reproduce thousands of years of paleoclimate.[11] If you store the data from a state-of-the-art GCM every five minutes, they can produce tens of terabytes per model year.

Another aspect of the models' complexity is their extreme "fuzzy modularity" (Lenhard and Winsberg 2010). In general, a modern state-of-the-art climate model is a model with a theoretical core that is surrounded and supplemented by various sub-models that themselves have grown into complex entities. The interaction of all of them determines the dynamics and these interactions are themselves quite complex. The coupling of atmospheric and oceanic circulation models, for example, is recognized as one of the milestones of climate modeling (leading to so-called coupled general circulation models). Both components had their independent modeling history, including an independent calibration of their respective model performance. Putting them together was a difficult task because the two sub-models now interfered dynamically with each other.[12]

Today, atmospheric GCMs have lost their central place and given way to a deliberately modular architecture of coupled models that comprise a number of highly interactive sub-models, like atmosphere, oceans, or ice-cover. In this architecture the single models act (ideally!) as interchangeable modules.[13] This marks a turn from one physical core – the fundamental equations of atmospheric circulation dynamics – to a more networked picture of interacting models from different disciplines (cf. Küppers and Lenhard 2006).

In sum, climate models are made up of a variety of modules and sub-models. There is a module for the general circulation of the atmosphere, a

---

[11] All of the above claims and quotations come from Dunne (2006).

[12] For an account of the controversies around early coupling, see Shackley et al. (1998); for a brief history of modeling advances, see Weart (2010).

[13] As, for example, in the Earth System Modeling Framework ("Earth System Modeling Framework" 2016)

module for cloud formation, for the dynamics of sea and land ice, for effects of vegetation and many more. Each of them, in turn, includes a mixture of principled science and parameterizations. And it is the interaction of these components that brings about the overall observable dynamics in simulation runs. The results of these modules are not first gathered independently and then only after that synthesized. Rather, data are continuously exchanged between all modules during the runtime of the simulation.[14] The overall dynamics of one global climate model is the complex result of the interaction of the modules – not the interaction of the results of the modules. This is why I modify the word "modularity" with the warning flag "fuzzy" when I talk about the modularity of climate models: due to interactivity and the phenomenon of balance of approximations, modularity does not break down a complex system into separately manageable pieces.

## 2) Distributed Epistemic Agency
Climate models reflect the work of hundreds of researchers working in different physical locations and at different times. They combine incredibly diverse kinds of expertise, including climatology, meteorology, atmospheric dynamics, atmospheric physics, atmospheric chemistry, solar physics, historical climatology, geophysics, geochemistry, geology, soil science, oceanography, glaciology, paleoclimatology, ecology, biogeography, biochemistry, computer science, mathematical and numerical modeling, time series analysis, etc.

Not only is epistemic agency in climate science distributed across space (the science behind model modules come from a variety of labs around the world) and domains of expertise, but also across time. No state-of-the-art, coupled atmosphere-ocean GCM (AOGCM) is literally built from the ground up in one short surveyable unit of time. They are assemblages of methods, modules, parameterization schemes, initial data packages, bits of code,[15] coupling schemes, etc., that have been built, tested, evaluated, and credentialed over years or even decades of work by climate scientists, mathematicians, and computer scientists of all stripes.

---

[14] In that sense, one can accurately describe them as parallel, rather than serial, models, in the sense discussed in Winsberg (2006).

[15] There has been a move, in recent years, to eliminate "legacy code" from climate models. Even though this may have been achieved in some models (this claim is sometimes made about CM2), it is worth noting that there is a large difference between coding a model from scratch and building it from scratch, i.e. devising and sanctioning from scratch all of the elements of a model.

No single person, indeed no group of people in any one place, at one time, or from any one field of expertise, is in any position to speak authoritatively about any AOGCM in its entirety.

### 3) Methodological Choices Are Generatively Entrenched

In our 2010 work, Johannes Lenhard and I argued that complex climate models acquire an intrinsically historical character and show path-dependency. The choices that modelers and programmers make at one time about how to solve particular problems of implementation have effects on what options will be available for solving problems that arise at a later time. And they will have effects on what strategies will succeed and fail. This feature of climate models, indeed, has lead climate scientists such as Smith (2002) and Palmer (2001) to worry that differences between models are concealed in code that cannot be closely investigated in practice.

Of course the modelers could – in principle – rework the entire code. The point is, however, that in even moderately complex cases, this is not a viable option for practical reasons. At best, this would be far too tedious and time-consuming. At worst, we would not even know how to proceed. So in the end, each step in the model-building process, and how successful it might be, could very well depend on the particular way previous steps were carried out – because the previous steps are unlikely to be completely disentangled and redone.

We argued that the best way to understand the historical nature of climate model optimization is in terms of a concept introduced by William Wimsatt in his recent book, viz., "generative entrenchment." Wimsatt explains generative entrenchment as follows: "A deeply generatively entrenched feature of a structure is one that has many other things depending on it because it has played a role in generating them"(Wimsatt 2007, p. 133). His discussion of this concept arises in the context of understanding how techniques from adaptive design function as "a way of increasing the reliability of structures built with unreliable components (...) Adaptive design is a layered organization of kludged adaptations acquired sequentially and assembled on the fly ..." (Wimsatt 2007, p. 133).

We further argued that "kludging" – the inelegant "botching together" (Clark 1987) of different pieces of code – plays a central role in the construction of complex models. Typically, a computational model is crafted over a long process of piecemeal mutual adjustments of its modules. For instance, to take a parameterization scheme from one model and make it work in the setting of another model requires mutual adjustments on the sides of both model and scheme code, often in the form of very pragmatic software-engineering measures that are not related to principled considerations.

When modifications are made to a complex model, and are shown to improve model performance, there is often a mixture of principled and unprincipled steps in the modification. When some new elements are added to a model, and improve model performance, it is often impossible to know if this happens because what has been added has goodness-of-fit on its own, or merely because, in combination with the rest of the model, what is achieved on balance is an improvement. Each successive modification of the model, then, has to not only be internally sound, but also to work harmoniously with previous components, and with the ways that they were kludged together. Finally, we argued that this generative entrenchment leads to an analytical impenetrability of climate models; we have been unable, and are likely to continue to be unable, to attribute all, or perhaps even most, of the various sources of their successes and failures to their internal modeling assumptions.

This last claim should be clarified to avoid misunderstanding. As we have already emphasized, different models perform better under certain conditions than others. But if model A performs better at making predictions on condition A*, and model B performs better under condition B*, then optimistically, one might hope that a hybrid model – one that contained some features of model A and some features of model B – would perform well under both sets of conditions. But what would such a hybrid model look like?

Ideally, to answer that question, one would like to attribute the success of each of the models A and B to the success of particular ones of their sub-models – or components. One might hope, for example, that a GCM that is particularly good at prediction of precipitation is one that has, in some suitably generalizable sense, a particularly good rain module. We called success in such an endeavor, the process of teasing apart the sources of success and failure of a simulation, "analytic understanding" of a global model. We would say that one has such understanding precisely when one is able to identify the extent to which each of the sub-models of a global model is contributing to its various successes and failures.

Unfortunately, analytic understanding is extremely hard to achieve in this context. The complexity of interaction between the modules of the simulation is so severe, as is the degree to which balances of approximation play an important role, that it becomes impossible to independently assess the merits or shortcomings of each submodel. One cannot trace back the effects of assumptions because the tracks get covered during the kludging together of complex interactions. This is what we called "analytic impenetrability" (AI)

(Lenhard and Winsberg 2010, p. 261). But AI makes epistemically inscrutable the effects on the success and failure of a global model of past methodological assumptions that are generatively entrenched.

To summarize the above points: climate modelers make choices that depend on the predictive tasks that they choose to prioritize; when they build new models, they build off old ones, and can never undo all the past choices that went into the older models on which they are building, and they are rarely in a good position to know exactly what all of those past choices were, and what effects they have had on the current model they are building. We'll see why all of that matters in the next section.

## 9.5 Models, Values, and Probabilities

Let's now step back from all of that and try to reconstruct the argument in a way that is both more abstract, and more carefully formulated in the language of Bayesian inference. According to the Bayesian view sketched earlier, the job of scientists is not to accept or reject hypotheses, but to determine the probabilities of those hypotheses in light of current information, K. The Bayesian defender of the value-free ideal (following Jeffrey) claims that this determination can and should be done in a value-free way in accordance with Bayes' rule, which provides a formula for updating the probability assigned to H in light of a new piece of evidence, e: $p(H|e) = p(H) \times p(e|H) / p(e)$. But we should make more explicit the role that background knowledge plays in Bayesian epistemology by putting background knowledge (**B**) into our expression of Bayes' rule. Bayes' rule then becomes: $p(H|e\&\mathbf{B}) = p(H|\mathbf{B}) \times p(e|H\&\mathbf{B}) / p(e|\mathbf{B})$. The combination of the new piece of evidence, e, and the previously held background information, **B**, together constitute **K**, the current information.[16] That is, $p(H|e\&\mathbf{B}) = p(H|\mathbf{K})$.

As I noted earlier, the defender of the value-free ideal is not claiming that the probability assigned to H is fully independent of non-epistemic values. After all, the contents of **K** will be value-dependent in the sense noted at the end of the last section: the background information **B** and new evidence e that agents have at a given time depend on what they considered important enough to investigate.[17] The claim I want to challenge here is

---

[16] Once the updating is performed, K then becomes B for purposes of updating in light of the next new piece of evidence.

[17] I argued earlier, following Wilholt, that this is *already* an important way in which social and ethical values play a role in hypothesis appraisal, and that it is not as "external" as many claim. But we are setting that point aside for now.

one that the Bayesian defender of the value-free ideal *is* keen to make: that social and ethical values need not and should not play a role in determining *the extent to which, given* **B***, the probability assigned to H should be increased or decreased in light of* e.

For ideal Bayesian agents, following this advice would be no problem: these agents would not be facing any of the challenges we detailed above – they would not be dealing with models that are analytically impenetrable, generatively entrenched, fuzzily modular, and constructed by a large, distributed group of scientists. But in the more complex epistemic situations that often arise in climate modeling, scientists trying to implement a Bayesian approach – and thereby trying to estimate the extent to which e, in conjunction with **B**, supports H – often *will* face these challenges, and hence will be influenced by non-epistemic values that affected earlier model-building choices.

In these cases, scientists' best attempts at estimating p(H|e&B) will often involve estimating p(H|e&B') instead, where **B'** replaces some of the claims in **B** with a computationally tractable scientific model, M, or set of models **M**, of the system or phenomenon under investigation. The model(s) will include some simplifications, idealizations, omissions and/ or other distortions relative to the claims in **B** it replaces. Thus, in practice, scientists' best attempts at estimating p(H|e&B) sometimes involve estimating p(H|e&B') instead, where **B'** is a distortion of **B** – eliminating some of the claims that are in B and replacing others with a computationally tractable scientific model, **M**. The distortions relative to **B** that scientists are willing to tolerate when developing **M**, however, will depend in part on the purposes and priorities of their investigations, as well as the purposes and priorities that shaped any earlier layers of the model's development.

The problem for BRAIR should now be clear. In practice, scientists' best attempts at estimating p(H|e&B) = p(H|K) sometimes involve estimating p(H|e&B') instead, where **B'** includes only a subset of the claims in **B** and replaces others with a computationally tractable scientific model **M**. But, as we've discussed, the distortions relative to **B** that scientists are willing to tolerate when developing **M** will depend on the purposes and priorities of their investigations and of the investigations of the model-builders on whose shoulders they stand. A clear example of this in climate science is the choice of a parameterization scheme, and values for the associated parameters, based on a assessment of "model skill" vis-à-vis a particular set of prediction tasks. The purposes and priorities of models clearly reflect social values; with different social values, some of the ways in which **M** (and thus **B'**) deviates from **B** would likely be different, too. Thus, in practice, social values sometimes do exert an influence as scientists attempt to estimate p(H|e&**B**).

*In principle*, of course, scientists could, for any model, correct for the ways in which the background knowledge embedded in the model, **M**, deviates from that in **B**. The problem is that when **M** is complex, non-linear, and motley, and is the product of many layers of model development, and the product of a wide array of disciplinary areas of expertise, it is nearly impossible for today's experts to perform such a calculation. For one thing, it may not be clear to a given group of scientists what all of the distortions of **M** relative to existing knowledge **B** are, nor what the effects on the results are; this could be because they didn't build the model or because the model is complicated enough that not all effects of choices are easy to anticipate, or both.

To be clear, I am not trying to be "model imperialist" about climate science – I do not assume that models embody the entirety of knowledge in climate science. Clearly, there are important sources of information about the future of the climate beyond complex simulation models. As we have discussed, these include, for instance: basic physical theory; simpler models, such as energy-balance and radiative-convective models; observations, including paleoclimate data; and knowledge of some of the ways in which actual models deviate from their ideal counterparts and from available observations. Nevertheless, this additional information does not necessarily make it easier to arrive at a confident and precise estimate of p(H|**K**) that is free from the influence of social values in the way that BRAIR envisions. On the contrary, interdependencies among different putative sources of information are often unclear (e.g. the information obtained from simpler and more complex models is not independent, but how does it relate?), and in any case these additional sources often fail to shed much light on the myriad ways in which past choices in model development, including choices influenced by social values, have shaped results. In practice, it is often infeasible to adjust model-based estimates to arrive at confident and precise estimates of p(H|**K**).

## 9.6 Imprecise Probabilities

It's worth returning, now, to a feature of the probabilities that are provided by the IPCC that we discussed in Chapter 6: They tend not to give precise probabilities to climate hypotheses. Rather, they tend to describe them as "likely," "very likely," "extremely likely," and so on; in a way that is meant to be decoded by the table at the end of section 6.1. Given what we have said so far in this chapter, it should no longer come as a surprise (if it ever did) that climate scientists should prefer to deliver coarse-grained ranges of probabilities rather than precise ones. When scientists cannot be confident that their estimate of $p(H|e\&\mathbf{B}')$ has brought them adequately close to $p(H|\mathbf{K})$, they are in a situation where their uncertainty about H is "deeper" than precise probabilities would imply.[18] Put differently, while the scientists might believe that an ideal agent would assign a single-valued probability to H given $\mathbf{K}$, they might recognize that as non-ideal agents, they themselves are uncertain about what that probability is. In a case like that, they might believe their uncertainty will be more accurately represented with imprecise probabilities (i.e. interval probabilities) or something even coarser. The same is true, of course, if scientists determine that $\mathbf{K}$ itself simply does not warrant assigning a single-valued probability to H. This might happen, for instance, if H pertains to the value of variable X, and $\mathbf{K}$ includes the claim that the mechanisms that control the value of X are rather poorly understood. I think this is the right way to understand why (or at least one reason why) the IPCC provides coarse-grained ranges of probabilities.

Given the topic of the chapter, we might then wonder if moving from precise probabilities assignments to climate hypotheses to ranges of probabilities can dampen the influence that social values, via modeling, have on conclusions reached.[19] To give a very simple example, suppose that a methodology for estimating $p(H|\mathbf{K})$ via model $M_1$ gives a result equal to $p_1$, while using $M_2$ it gives a result equal to $p_2$. Whatever sets of social values in fact shaped the development of $M_1$ and $M_2$ – let us suppose that they are somewhat different sets of values – a declaration that $p(H|\mathbf{K})$ is somewhere in the interval $(p_1, p_2)$ would be neutral with respect to the choice between those sets of values.

Nonetheless, while it's true that the move to coarser uncertainty estimates has the potential to diminish the influence that social values, via a model or set of models $\{\mathbf{M}\}$, exert on the conclusions reached, it is not clear how "escapable" the influence of social values ultimately is using this strategy. There are two reasons for this.

The first reason has to do with the effect that working with a set of models $\{\mathbf{M}\}$ has on one's thinking. There is some process that IPCC scientists go through to arrive at some conclusion. Presumably that process involved

---

[18] Of course, this is not the same thing as having lower credence in H.
[19] Parker (2014) makes this suggestion.

consideration of, among other things, the limitations of the models used to generate the projections, insofar as those limitations are understood. Perhaps they recognize, for example, that cloud feedbacks are represented rather poorly. The challenge is to reason about how the projections might have been different if those feedbacks had been accurately represented (relative to background knowledge **B**) and if other distortions relative to **B** were similarly removed. Unlike when single-valued probabilities are sought, this reasoning can be rather coarse-grained. I would argue, however, that this reasoning will still be shaped by scientists' past experience with the models that happen to be on offer, e.g. by the range of results produced by models that incorporate different representations of the feedbacks in question. After all, models are often used in the study of complex systems in part because it is very difficult to reason about the system without them. This at least raises questions about the extent to which the social values that shaped the development of {**M**} can ultimately be escaped.

The second reason is that a move to coarser uncertainty estimates makes room for a whole new set of social values to enter into the mix. An IPCC example is instructive. The IPCC scientists reported "high confidence" in the conclusion that warming greater than 2°C was "likely" under RCP6.0. The "likely" range corresponds to an interval probability assignment of (0.66, 1.0). But the scientists could also have reported a wider probability interval, such as (0.5, 1.0), corresponding to "more likely than not," with even higher confidence, e.g. "very high confidence." Or perhaps they could have reported a narrower interval, such as (0.9, 1.0), corresponding to "very likely," but with less confidence. The question is: what determines which of these representations of uncertainty is communicated? (We assume here that the representations are all consistent with one another.) Without speculating regarding the IPCC example, it seems clear that at least sometimes it is a consideration of the likely applications of an uncertainty report that guide the choice between a wider and more confident report and a narrower and somewhat less confident report. Perhaps a narrower, even if somewhat less confident interval is thought to be more useful for policy makers. In such cases, social values are once again playing a role.

One point worth noting in the context is that by providing fixed intervals of coarse-grained probability that climate scientists have to choose from, the IPPC is in fact removing one degree of value-laden freedom. If it is a value-laden process for me to decide whether I should report, with high confidence, that pr(H) is between .90 and .95, or report, with lower confidence, that pr(H) is between .90 and .91, then by forcing me to use the (.90, .95) bin, the IPCC has taken away a dimension of value-ladenness from me and forced me to follow a conventional path. In most cases, we should view this as a good thing, and as a strong advantage of the imposition of the chart we first encountered in section 6.1.

## 9.7 Values and Biases

There is one final remark worth making about the claim that evidence evaluation and scientific conclusions are value-laden. This is not at all the same as the claim that scientific conclusions are reached in a way that is systematically biased. I am not defending any sort of claim about climate science ever being *biased*. I believe this is a common misunderstanding of what the science and values literature claims, including by some climate scientists. Consider, for example, the following passage from "A Practical Philosophy of Complex Climate Modelling" by Gavin Schmidt and Steven Sherwood:

> Winsberg and Douglas suggest that, if given a choice between two different parameterisations with similar overall contributions to model skill, a modeler might choose one that yields predictions that might err in a specific way, for instance, producing under- or over-estimates of some feature matching their prior preference for avoiding false negatives over false positives. However, there are several reasons why such suggestions can be dismissed. First, it would require enormous study and expense to work out how to configure a model to produce preferred predictions (or predictions with a preferred sign of error), with little evident benefit to the modelling group, and enormous risks to their credibility if it became known that they had done so (which would be hard to avoid given the expense). The status of climate simulations in the public policy environment is already highly contested, and scientific credibility of the climate model development process is a frequent topic of debate. Second, models are now used to predict so many things under a wide range of applications that it is hard to imagine a modelling group or individual modeller settling on one particular outcome (or bias) to aim for, given the (indeterminate) impact this would inevitably have on other predictions. Third and most importantly, this activity would interfere and compete with the far stronger imperative to achieve the best skill possible against observable metrics in today's (or past) climate, based on modelling decisions that are scientifically and practically defensible. Specifically, the effort required to assess the varying sensitivity of the model to multiple drivers as a function of the two parameterisations would be much better spent varying the parameterisations more finely and assessing improvements to climatological skill. (Schmidt and Sherwood 2015 , p. 155)

In reality, the facts that Schmidt and Sherwood cite are excellent grist for my mill. When complex models are the source of the values, there is little reason to think that the influence of values will be biasing in line with the social values that shaped past decisions. In some cases indeed, it is not entirely clear what such bias could consist in, since the influence of social values via predictive preferences in modeling do not even take the right form for doing this. Unlike in the context of inductive risk, they are not influences that make it more likely that one conclusion rather than another will be reached.

In fact, even though many modeling choices are choices to prioritize some modeling tasks and variables over others, the effect that those choices will have in the long run may or may not be to make the model better at the tasks and variables that were the basis for the prioritization. It all depends on how they get stacked up together. And the same is the case with inductive risk preferences. In complex models, there is no simple additive formula for the compound effect of multiple modeling choices, and later choices in model development can "undo" the work done by a past choice. In an earlier paper, I said that values are "in the nooks and crannies" of climate models, to emphasize this point. The long run and overall effects of a series of modeling choices are impossible to judge. Indeed, if it were possible to systematically judge the long run and overall effects of these modeling choices, then the BRAIR reply would be right! The smart climate scientist would just account for all these effects in her estimation of the probabilities. The important point that I want to make here is that while there is likely to be a dependence of the final outcome on value-motivated past choices, it is overwhelmingly unlikely that those effects will be systematic or predictable. And so when Schmidt and Sherwood say, "it would require enormous study and expense to work out how to configure a model to produce preferred predictions," I say, "Yes! This is exactly what the 'nooks and crannies' claim was meant to convey!"[20]

## 9.8 Conclusion

Philosophers sometimes worry that allowing a role for values in the evaluation of scientific hypotheses, models, or theories will lead to "wishful thinking." This is the worry that scientists will be able to choose which claims they want to be true, and in so doing, make it more likely that they will be so evaluated. But insofar as we only accept arguments for values

---

[20] Wendy Parker and I have a bit more to say about the relationship between values and biases in Parker and Winsberg (2017).

whose influence cannot be corrected for by the Bayesian, we will not have to have this worry.

Suppose at several uncertain methodological decision points I choose the options that I have quite good reason to believe will push the value of a particular output variable higher rather than lower. It's not hard to imagine that I could understand how to do this for at least some output variables – otherwise people wouldn't be able to tune models at all. In such a case I can push the result for the variable in the direction I want. But if I have done this, and I am following good scientific practice, then *I should adjust my probabilities accordingly*. If my models predict a central tendency for that variable to be, let's say, 5, then I should assign the probability of the true value of the variable being greater than 5 to be something less than 50 percent, because I know I've erred on the upside. If I don't make that adjustment, then you can accuse me of wishful thinking. But then your argument that I engaged in wishful thinking will ipso facto be an argument that I should have better adjusted my probabilities. If I can successfully reply that I have done the best job I can in adjusting my probabilities, then I have ipso facto given an argument that I have not engaged in wishful thinking.

In general, an argument that some scientist has exhibited wishful thinking ought to be, ipso facto, an argument that she *should* have adjusted her probability. It is therefore, ipso facto, an argument that the influence of her values *could have been corrected*. Conversely, an argument that some value influence was *uncorrectable* (by Bayesian methods) has to ipso facto also be an argument that wishful thinking could not have played a role.

## Suggestions for Further Reading

Rudner, Richard. 1953. "The Scientist Qua Scientist Makes Value Judgments." *Philosophy of Science*. One of the earliest articulations of the argument from inductive risk for values in science.

Jeffrey, Richard C. 1956. "Valuation and Acceptance of Scientific Hypotheses," *Philosophy of Science*. Jeffrey replies to Rudner by arguing that scientists should avoid accepting and rejecting hypotheses and should only assign them probabilities.

Douglas, Heather. 2000. "Inductive Risk and Values in Science," *Philosophy of Science*. The paper that brought the Rudner/Jeffrey debate back into philosophy of science. Douglas argues, pace Jeffrey, that value judgments need to be made prior to hypothesis acceptance and rejection.

Steel, Daniel. 2015. "Acceptance, Values, and Probability," *Studies in History and Philosophy of Science Part A*. Argues that on the right understanding of the role that credences play in science, we ought to think of scientists accepting or rejecting probability assignments, rather than simply coming to have them. On this view, Steel argues, acceptance and rejection of a probability assignment itself involves inductive risk.

Longino, Helen. 1995. "Cognitive and Non-Cognitive Values in Science: Rethinking the Dichotomy," in *Feminism, Science, and the Philosophy of Science*. Challenges the distinction, often appealed to in the science and values literature, between epistemic and non-epistemic values.

Parker, Wendy and Eric Winsberg. 2017. "Values and Evidence: How models Make a Difference," *European Journal for Philosophy of Science*.

# 10

## *Skill*

### 10.1 Introduction

Chapter 8 of Working Group 1's report in the AR4 is entitled "Climate Models and their Evaluation." It contains "FAQ 8.1" which asks: "How Reliable Are the Models Used to Make Projections of Future Climate Change?" The answer begins:

> There is considerable confidence that climate models provide credible quantitative estimates of future climate change, particularly at continental scales and above. This confidence comes from the foundation of the models in accepted physical principles and from their ability to reproduce observed features of current climate and past climate changes. Confidence in model estimates is higher for some climate variables (e.g., temperature) than for others (e.g., precipitation). Over several decades of development, models have consistently provided a robust and unambiguous picture of significant climate warming in response to increasing greenhouse gases. (AR4, WG1, 8.2, FAQ 8.1)

In the next few chapters, we will be primarily interested in the claim that "confidence comes from the foundation of the models in accepted physical principles and from their ability to reproduce observed features of current climate and past climate changes." In the next three chapters, we will try to take a more systematic and philosophical approach to understanding the methods that climate scientists use for evaluating what they often call the "skill" of their models. Climate scientists do not always mean the same thing when they speak of model skill, but in keeping with some of the conclusions we reached in Chapter 3, when we talk about evaluating climate models, or evaluating a climate model's skill, we will be referring to the activity of evaluating a model's adequacy for particular purposes. In particular, we will be especially interested in philosophical issues that arise in thinking about three sources of evidence we can have for models' skill: their "foundation ... in accepted physical principles"; their "ability to

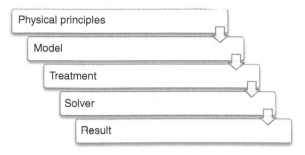

Figure 10.1 The "linear model" of computer simulation.

reproduce observed features of current climate and past climate changes"; and (quoting a little further down in the text), the extent to which they "provide[…] a robust and unambiguous picture."

## 10.2 Verification and Validation

Let's start with the first source of model confidence: their foundation in accepted physical principles. What, exactly, does that amount to? As we saw in Chapter 3, the core of all climate simulations is a set of five equations that govern the general circulation of the atmosphere which are, in turn, grounded in unassailable (in the relevant domains) physical principles like conservation of mass, Newton's laws, and the simple thermodynamics of heat transfer. How should we think about the role that these well-accepted principles play? One way of thinking about it is captured by what one might call the "linear model of computer simulation" (to emphasize how we move in a straight line from principles to results). This is closely related to what Nancy Cartwright has called "The vending machine view of theories." Since the linear model has implications for the epistemological significance of the fact that climate models are founded in accepted physical principles, we should scrutinize it here. The model is illustrated in Figure 10.1.

The idea is rather simple. We can illustrate it with a simple example: the pendulum. We begin with a theory that governs the phenomenon of interest. For a pendulum, that would be Newtonian mechanics. A basic understanding of the system suggests to us a model of the pendulum. Such a model would consist of an abstract description of the physical system (a point on the end of a massless string), and some differential equations provided by the theory that describe the evolution of the values of the variables in the abstract model. The treatment consists of assigning values to basic parameters – in this case the length of the string, the mass of the bob, and the acceleration of gravity – and assigning initial values to the variables, which in this case would be an initial value for the angle of the pendulum and its angular velocity. Next, model and treatment are combined to create a solver – a step-by-step computable algorithm that is designed to be an approximate, discrete substitute for the continuous differential equations of the model – and the solver is run on a computer to produce results.

Despite the simplicity of the example, this schema captures a great deal about many computer simulations of physical systems. Associated with that schema is a conception of the epistemology of simulation that is almost disarmingly simple and straightforward. The phrase associated with that idea is "Verification and Validation" or just "V&V" and it is built around the idea that there is a clean and simple distinction between the two "V"s that goes something like this:

- Verification: "the process of determining the extent to which the solutions generated by the computer simulation model approximate the real solutions to the original mathematical model equations."
- Validation: "the process of determining whether the original model equations are the correct or appropriate ones."

Alongside that distinction is a pair of claims about the relationship between the two distinguished processes:

- "Separability claim":
  o Validation and Verification are distinct activities that are carried out serially, and. give rise to distinct epistemological concerns that can be addressed on their own.
- "Disciplinarity claim":
  o Verification is a matter of *mathematics* and validation is a matter of *physics/chemistry/atmospheric science/etc.*

We can find these kinds of claims both in the scientific literature and the philosophical literature.

From the engineer Christopher Roy, for example, we get:

> Verification deals with mathematics and addresses the correctness of the numerical solution to a given model. Validation, on the other hand, deals with physics and addresses the appropriateness of the model in reproducing experimental data. Verification can be thought of as solving the chosen

equations correctly, while validation is choosing the correct equations in the first place. (Roy 2005, p. 133)

And from the philosophers Roman Frigg and Julian Reiss we get:

> We should distinguish two different notions of reliability here, answering two different questions. First, are the solutions that the computer provides close enough to the actual (but unavailable) solutions to be useful? ... *this is a purely mathematical question* and falls within the class of problems we have just mentioned. So, there is nothing new here from a philosophical point of view and *the question is indeed one of number crunching*. Second, do the computational models that are the basis of the simulations represent the target system correctly? That is, are the simulation results externally valid? This is a serious question, but one *that is independent of the first problem*, and one that equally arises in connection with models that do not involve intractable mathematics ... (Frigg and Reiss 2009, p. 594)

In what follows, I will be arguing that some of these claims are misleading when they are applied to the epistemology of climate modeling. It is certainly true that climate scientists do engage in activities that can be labeled verification activities or validation activities and that, indeed:

1) Verification and Validation are conceptually distinguishable sorts of activities, and
2) In the actual practice of climate simulation, there are activities carried out that can or should be understood as either directed primarily at verification or at validation[1].

But what I would like to make clear is that the general epistemological account of how simulation results are evaluated that goes by the expression "verification and validation" is very much tied to the separability claim and the disciplinarity claim. And neither of those claims are true of climate science and its credentialing activities.

The reason that climate scientists cannot genuinely verify and validate their models separately has to do with many of the features of climate simulation models we discussed in Chapter 4: the role of parameterizations, fuzzy modularity (see also Chapter 9), and tuning. All of these features of

---

[1] Verification-directed activities include things like software quality assurance techniques, consistency and convergence tests (roughly speaking: making sure that as you make space and time grid cells smaller and smaller, the results you get start to change less and less) and doing comparisons of simulation results with known solutions. Validation-directed activities including comparing your model's output to the output of a model that has stronger theoretical underpinning, etc.

climate simulation models render the "linear model" we discussed above inapplicable to climate modeling. Rather than undergoing "linear development," the development of simulation models in sciences that are as complex as climate science has more of a "life cycle."

The "life cycle" of a typical complex simulation model in climate science goes something like this.

1) Well-accepted physical principles suggest a model M1;
2) M1 is discretized and implemented in a computer simulation via scheme S1;
3) The output is compared to the real world. If the output is found to lack agreement with data, be unstable or otherwise unphysical, you proceed to:
4) Basic features of either the model or the discretization scheme are changed:
   o Either by parameterizing an effect that failed to be adequately captured, adding an effect that was not represented in the old model or by changing the way the underlying model is represented in the discrete algorithm;
   o The ways in which M1 and S1 can be modified are no longer suggested by well-accepted principles or theories, but come from other sources: physical intuition, phenomenology, local empirical findings, lore accumulated from parallel modeling successes, etc.;
5) M2 is discretized with S2;
6) Go back to 3, rinse, repeat;
7) Eventually, a model/discretization-scheme pair is found that can be sanctioned.

When a simulation model is developed in an iterative process like we see in Figure 10.2 – one in which model development and model sanctioning are occurring hand in hand, there can be no justification of the final model that is independent of its discretized implementation, and there can be no justification of the implementation that is independent of the model. In such a situation, the modelers have to settle on a model that they could never justify on a purely theoretically principled basis. There are, in the end, at best very weak theoretical arguments to be given for the ultimate choice of model. The ultimate justification of a set of model equations is that they work – that the final output of the simulation can be argued to be reliable (enough). There can be no independent "validation" of the model.

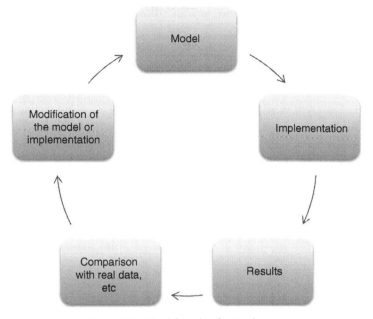

Figure 10.2 The life cycle of a simulation.

There are also, at best, weak mathematical arguments, weak in the sense that they only make the results seem plausible, rather than guaranteed, that can be given that the output of a simulation approximates the mathematical content of those models. This is because climate models are so complex, and rely so heavily on non-linear equations, that when these arguments are applied to the entirety of solution space, few guarantees can be offered. When models are sufficiently complex and non-linear, there can be no independent "verification" of the model.

When a climate model succeeds at passing whatever tests we subject it to, it might be because the underlying model is ideal and the algorithm in question finds solutions to that underlying model. Or it might be because of a "balance of approximations." This is likely the case when a model is deliberately tailored to counterbalance what are known to be limitations in the schemas used to transform the model into an algorithm. To the extent that climate modelers use parameter values to "tune" their models, particularly when they tune them to avoid radiation imbalances in the TOA (see Chapter 4.4) that are the result of errors elsewhere in the model, these models are explicitly believed to represent a balance of approximation. And when success is achieved in virtue of this kind of back-and-forth, trial-and-error piecemeal adjustment, it is hard to even know what it means to say that a model is separately verified and validated.

In fact, the results of climate simulations are sanctioned all at once: users try to maximize fidelity to theory, to mathematical rigor, to physical intuition, and to known empirical results and to minimize known computational pitfalls. But it is the *simultaneous confluence* of these efforts, rather than the establishment of each one separately, that ultimately gives us confidence that the results are adequate for any particular purpose.

## 10.3  An Alternative Picture

The right story about why we should have "confidence that climate models provide credible quantitative estimates of future climate change" had better reflect this fact, and the V&V story does not. In other work, I have argued that the reasons simulation modelers give for having confidence in their results are complex and motley. A big part of the story I pursued in that work involved recognizing the "craft" of simulation. The move from basic accepted physical principles to the working model that is depicted in Figure 10.2 is usually full of approximation, idealization, parameterization, and other techniques of transformation. Simulation models are most confidence-inspiring, I argued, when these techniques are deployed by practitioners that are well-acquainted with the techniques' strengths and weaknesses, and when they are knowledgeable not only about which physical principles are well-accepted, but also about which model-construction techniques are well-credentialed – and for what purposes and in what contexts. I argued that these kinds of techniques "have a life of their own" (Winsberg 2003, p. 121). This means that, "they carry with them their own history of prior successes and accomplishments, and, when properly used, they can bring to the table independent warrant for belief in the models they are used to build" (Winsberg 2003, p. 122).

Not everyone will find a story like this entirely satisfying. William Goodwin, in particular, has expressed the worry that this kind of story has "no normative force." He worries that it:

> does nothing to establish the legitimacy of such techniques, but instead reports that their acceptance as legitimate is bound up with their detailed history of use. If this description is right, however, then it suggests that any convincing normative accounts of why particular simulation techniques are legitimate, or reliable, will be very local and detailed. Perhaps a simulation modeler could explain to his peers why it was legitimate and rational to use a certain approximation technique to solve a particular problem, but this

explanation would – if Winsberg is right – appeal to very context specific reasons and particular features of the history of the technique. This does not suggest that such techniques are illegitimate or unjustified, but it does make it unlikely that the legitimacy of such techniques can be reconstructed, or rationalized, in terms of generally recognized, ahistorical evidential norms. As is often the case, it may be difficult or impossible to "step back" from the practice of global climate modeling and argue, independently, for its legitimacy; and this makes the absolute strategy for establishing the predictive reliability of global climate models unlikely to be successful. (Goodwin 2015, pp. 342–343)

I have to confess that, unfortunately, I think Goodwin is right about the shortcomings of my account. It means that "a philosopher who hopes to address the epistemological concerns that have been raised about the reliability of climate models" isn't going to be able to tell a normatively grounded story that will secure the unassailable reliability of the results of climate modeling. But I still think the account is more-or-less right. I agree that it would be great if philosophers could do the above, but in fact the track record of philosophers in normatively grounding any kind of knowledge has been less than terrific. I think that when we look on the work of those who are in the business of modeling highly complex non-linear systems, the best we are ever going to be able to do is to arrive at a situation where "a simulation modeler could explain to his peers why it was legitimate and rational to use a certain approximation technique to solve a particular problem" by appealing to "very context specific reasons and particular features." The more time that I have spent talking to practitioners, the more my belief in this view has been reinforced. The responses that I hear to what prima facie sounded like plausible skeptical worries about particular conclusions often appeal to fairly esoteric details of the situation. We will come back to this point in much more detail in Chapter 13, but my view is that philosophers do better to paint a picture in which we urge trust in the consensus of the scientific community, based on features of that community's social organization, then to try to provide a normative framework from which we can demonstrate the reliability (or its absence) of such-and-such modeling result.

Goodwin is also somewhat pessimistic about the potential for giving an "absolute" philosophical argument for the reliability of global climate models – that is, one that would ground their reliability in an ahistorical normative framework. Instead, he prefers a "relative" strategy, one where the philosopher shows that climate simulation models are on a par, epistemically, with scientific practice that is not typically called into question. Such a strategy "depends upon finding a set of predictions that are broadly recognized to be reliable and which can be plausibly argued to be relevantly similar to the sorts of quantitative predictions generated by GCMs" (Goodwin 2015, p. 343).

Such a strategy is fine – it is certainly helpful to remind people that climate models are not in any deep and interesting way sui generis[2] – but tight analogies of the kind that Goodwin is looking for might be hard to come by. Given any other "relevantly similar" set of predictions one tries to use in order to "relatively" underwrite the reliability of climate models, there are likely to be relevant differences, alongside the relevant similarities. Certainly, if one looks to engineering contexts for instances of applied science, the ways in which models are sanctioned will display some salient differences with respect to climate science – particularly with respect to the extent to which each respective discipline seeks to apply techniques in ways that might turn out to be outside of the domain under which they have been directly tested. In some engineering contexts, more-over, the V&V framework is actually more or less applicable. Some engineers have made the point to me that in highly sensitive applications, such as simulations used to assure the safety of the nuclear stockpile, it is fundamental and important that the V&V framework apply (Oberkampf and Roy 2010).

We should also be careful about trying to secure more confidence in climate simulations than is possible *or* needed. We do not need to secure the reliability full-stop of climate models. Goodwin *is* correct to have no patience for philosophers and others who want to entirely undermine all confidence in the reliability of the simulation models and who insist that "basic climate science and the accumulated climate data are enough to establish the reality of climate change, diagnose its source(s), and differentiate between policy options (e.g. would halving our emissions make any difference?)" (Goodwin 2015, p. 341, fn 5).

But for exactly the same reason that Goodwin thinks philosophers should avoid making their own determinations about how much evidential weight can be placed on basic science and data, I think philosophers should also be careful about making their own determinations of how much and exactly what forms of model reliability are needed to underwrite the well-accepted claims of climate science. One-line quotations from the IPCC, moreover, cannot be used to establish the truth about

---

[2] Though of course this will be no help in responding to folks who think that all non-linear modeling is "embroiled in confusion."

this. Scientists, in any case, are not epistemologists, and they are not necessarily experts on exactly what foundation their beliefs rest – even if they are the best experts regarding what *should be believed*. Those are not the same question. The epistemic role that climate simulations play in securing confidence in the IPCC's most important claims (for example) is more complicated than some of these one-liners make it out to be, as much as it is more complicated than the "basic science" proponents make it out to be.

What really needs to be secured is confidence in the claim that climate models make reliable projections or predictions, whatever exactly that would mean. What needs to be secured is confidence in the contribution that simulations make to our overall understanding of the climate system, which *in turn* underwrites our confidence in the various probabilistic claims that the IPCC is committed to. And that is no simple matter to tease apart. It is certainly not the same as having confidence in the predictive reliability of the simulations *tout court* (again: whatever exactly that would mean). Our most important climate hypotheses cannot be secured without any help from simulations, but they also aren't naively read off of the results of those simulations. Simulations can, for example, gradually and in piecemeal fashion contribute to our process understanding of the climate system (see Box 11.1). As such, they can contribute to our confidence in various hypotheses in a mediated way, rather than it being a case of "we have confidence in this or that hypothesis because it is a straightforward result in our simulations." An excellent example of this is the way in which simulations contribute to our knowledge of equilibrium climate sensitivity, which is discussed in Chapter 12. Simulations can also contribute to our knowledge of transient climate sensitivity by providing inputs for values in back-of-the-envelope dynamical models which would otherwise not be useful. But each time such a simulation does contribute to this tapestry of evidence that supports well-accepted claims, it is done vis-à-vis the "very context specific reasons and particular features" that Goodwin, understandably, finds somewhat unsatisfying.

## 10.4 Tuning Revisited

Whatever turns out to be the true answer to the question of how reliable climate simulation models are, there is little doubt that whatever degree of reliability they have, they get at least some "from their ability to reproduce observed features of current climate and past climate changes." But given that these same simulation models are tuned against existing data concerning those same observed features, this raises an obvious question.

How much reliability can we infer from the ability of a simulation to reproduce data against which it was tuned? A debate about this question, which is sometimes characterized as the question of whether or not "double counting" is permitted in model evaluation, has arisen in the philosophy of science. Katie Steele and Charlotte Werndl (2013) have argued that there is nothing at all wrong with using the same data to test a model as one used to tune it. They claim, in fact, that this practice is nothing other than the ordinary practice of using evidence to test a hypothesis. They argue that this conclusion is a simple consequence of the Bayesian approach to hypothesis testing (see Chapter 8). Mathias Frisch, on the other hand, has argued that Steele and Werndl overstate what the Bayesian framework commits us to (Frisch 2015).[3] As we learned in Chapter 4, tuning is an important part of climate model construction. We should therefore say a few things about this debate.

Let's begin by canvassing some fairly typical remarks about this issue by climate scientists.

> Including quantities in model evaluation that were targeted by tuning is of little value. Evaluating models based on their ability to represent the TOA radiation balance usually reflects how closely the models were tuned to that particular target, rather than the models intrinsic qualities. (Mauritsen et al. 2012, quoted in Frisch 2015)

> Model tuning directly influences the evaluation of climate models, as the quantities that are tuned cannot be used in model evaluation. Quantities closely related to those tuned will provide only weak tests of model performance. Nonetheless, by focusing on those quantities not generally involved in model tuning while discounting metrics clearly related to it, it is possible to gain insight into model performance. Model quality is tested most rigorously through the concurrent use of many model quantities, evaluation techniques, and performance metrics that together cover a wide range of emergent (or un-tuned) model behaviour. (IPCC, AR5, WG1, Box 9.1, quoted in Frisch 2015)

> If the model has been tuned to give a good representation of a particular observed quantity, the agreement with that observation cannot be used to build confidence in that model. (IPCC, AR4, WG1, p. 596)

> It is incumbent upon those who develop models to know how they have (and have not) been tuned, in order to avoid inappropriate conclusions from successful tests, though the literature has historically been a little opaque on this topic. Though perhaps an obvious point, characteristics (or metrics) that

---

[3] Much of the rest of this section follows Frisch (2015). The last part, in which I square off some of Frisch's claims against the stronger claims of climate scientists, is of course entirely my own.

are used to explicitly tune a model or its inputs should not also be used to evaluating the model – this would be a form of 'double counting'. A recent paper argued the opposite, that in fact some kinds of 'double counting' are both permissible and practiced (Steele and Werndl 2013). On closer inspection though, both examples of 'double counting' addressed in that paper are simple versions of parameter tuning or model selection with no evaluation beyond the fitting procedure. The authors describe this as 'relative confirmation' (among models), but in our opinion that is irrelevant to assessments of model predictive skill which is the point that we are concerned with here. Specifically, Steel and Werndl ignore the fact that results that are predicted "out-of-sample" demonstrate more useful skill than results that are tuned for (or accommodated). (Schmidt and Sherwood 2015, p. 9)

Frisch points out that there are two readings of the concern. The IPCC quotation and the Schmidt and Sherwood quotation (which Frisch does not canvass) support Frisch's "stronger claim" that "data used in tuning simply cannot be used in model evaluation or in the confirmation of models at all" (Frisch 2015, p. 190), or as Schmidt and Sherwood put it, that data that have been used to tune a model are "irrelevant to assessments of model predictive skill" (Schmidt and Sherwood 2015, see above). Frisch's "weaker claim" is that "data not used in tuning can be more highly confirmatory of a model or provide a superior test in model evaluation than data used in tuning" (Frisch 2015, p. 190.) Frisch argues, pace Steel and Werndl, for the weaker claim, while Schmidt and Sherwood (obviously) argue for what Frisch calls the stronger claim. I will suggest below that the view of Schmidt and Sherwood is the correct view to hold, though I will also point out that their view is not *strictly* stronger than Frisch's "weaker claim."

We already discussed tuning a little bit in Chapter 4.4. In that section, we characterized tuning as "the process of adjusting a few key parameter values (associated with the sub-grid parameterizations discussed above) in order to get the behavior of the model *as a whole* to behave in an acceptable way." Here it is worth highlighting another feature of tuning that is especially relevant to assessing the permissibility of "double counting." First, "tuned" parameters are generally not adjusted to what we would ordinarily think of as their "correct" values. This is a tricky point because parameter values in climate models often do not correspond to a value of anything in the real world. But there are clearly cases in which parameter values are picked with the explicit goal of compensating for structural errors in other parts of the model in which those parameters live. As we mentioned in Chapter 4.4, cloud parameters are used in order to correct for a structural error that persistently plagues global climate models: they "leak radiation" in a way that makes the top-of-the-atmosphere (TOA) energy balance incorrect. But because we do not know how to correct the structural error that leads to the TOA imbalance, we instead "fix it" by adjusting the values associated with another structural error: the representation of cloud dynamics in the model.

Another example canvassed by Frisch (2015, p. 175) is the tuning of the autoconversion threshold radius in the CMIP5 version of the GFDL CM3 model. This was done in order to achieve the best possible fit with twentieth-century temperature data. The best value that was found was determined to be one that was almost surely not realistic. As the authors of the study reporting that tuning note: "This might indicate the presence of compensating model errors" (Golaz, Horowitz and Levy 2013, p. 2249, quoted in Frisch 2015, p. 175).

It would seem obvious that, since most climate models are tuned *to get the TOA energy balance right*, the fact *that they get the TOA balance right* is not evidence of their skill at doing anything else. Given that the GFDL CM3 is tuned to achieve the best possible fit with twentieth-century data better than the CM2, the fact that CM3 does a better job of representing the twentieth-century data better than the CM2, is not, *by itself*, evidence that the CM3 has more skill than the CM2. Why, then, do Steele and Werndl dispute this?

Their argument is made in the context of Bayesian confirmation theory (see Chapter 8). Bayesian confirmation theory tells us how to calculate the "posterior probability" $P_f(S)$ of a hypothesis from its "prior probability" $P(S)$ and from the "likelihood," the probability of E given S, or $P(E|S)$:

$$P_f(S) = \frac{P(E|S)P(S)}{P(E)}$$

This is what is known as "Bayes' rule". It is a simple consequence that the degree to which some new evidence E confirms some hypothesis S1 relative to another hypothesis

$$\frac{Pf(S1)/P(S1)}{Pf(S2)/P(S2)}$$

is given by

$$\frac{P(E|S_1)}{P(E|S_2)}$$

This mathematical reasoning, moreover, is independent of whether we consider the prior and posterior probabilities to be probabilities of the model themselves, of the hypothesis that the model is "valid" or "confirmed" or whatever, or of the hypothesis that the model has some degree of skill. It's a simple consequence of the assumption that Bayes' rule governs how we should update our beliefs in the light of new evidence – and few if any philosophers of science doubt this assumption. Importantly, moreover, it looks as if it shows that the relative degree to which two hypotheses are confirmed by some body of evidence depends only on the likelihoods, and the likelihoods do not have anything to do with the biographies of the hypotheses: they have nothing at all to do with whether the people who constructed the hypotheses (or the models that the hypotheses are about) used the evidence in question to construct them. So it looks as if, by the light of Bayes' rule, there is no difference whatsoever between data used to tune a model, on the one hand, and other data, on the other, when it comes to the degree to which those data can evidentially support the model.

So what, if anything, went wrong?

One thing to note is that the version of Bayes' rule that we gave above is missing one element which is very frequently suppressed but is actually crucial to include in certain contexts: the background knowledge. After all, the probabilities in Bayesian inference are best thought of as something like degrees of belief (see Chapter 6). And it doesn't really make sense to talk about my prior degree of belief in a hypothesis, or of a likelihood, in a vacuum. These degrees of belief are always conditional on my background knowledge. Strictly speaking, the correct version of Bayes' rule should have all of the probabilities on both sides of the equality given as conditional on my background knowledge, which includes all of what I believe.

$$P_f\left(S|B\right) = \frac{P\left(E|S \& B\right)P(S|B)}{P(E|B)}$$

The problem is that on this (fully specified) version of Bayes' theorem, E is part of, and is thus a logical consequence of, our background knowledge B. But if E is logically entailed by B, then necessarily $P(E|B)=1$. By similar reasoning, $P(E|S\&B)=1$ as well. And it follows from this that the posterior probability of any such hypothesis is exactly equal to its prior probability. These results obtain for any E that represents evidence that was known at the time of model construction. And that means that no evidence that is known at the time of model construction shifts any of our degrees of belief about the model.

This is what is known in the literature on Bayesian inference as the "problem of old evidence." It is called the "problem" of old evidence because many have the intuition that this result is not right – that at least in some cases, evidence that was indeed known at the time of the construction of a hypothesis can in fact be used to confirm it. One example that is taken to be a canonical case of this is the relationship between the precession of the perihelion of Mercury and the theory of general relativity. Even though a detailed description of the precession was widely known prior to Einstein's formulation of the theory of general relativity, most people believe that this piece of evidence – the perihelion of Mercury data (PMD) – did allow physicists to appropriately raise their degrees of belief in GR.

And so of course the literature contains several possible solutions to the problem.

1) One possible solution is to divide our background knowledge into the part that is used in the construction of the model or hypothesis and the rest, and to call the remainder of the background knowledge "use-novel." On this account, any part of B that is not used in model construction is allowed to be left out of the correct formulation of Bayes' rule. Let B' be the set of background beliefs in B that are not "use-novel" and set the new version of Bayes' to be

$$P_f\left(S|B'\right) = \frac{P\left(E|S \,\&\, B'\right)P(S \,|\, B')}{P(E \,|\, B')}$$

In this proposed solution, because PMD is not part of B', the PMD can be used to confirm GR precisely because it is use-novel with respect to GR. On this view, however, since data used to tune a climate model are not use-novel, they cannot be used to confirm the model or any hypothesis regarding the model.

2) Another Bayesian solution to the problem of old evidence is to deny that Bayesian reasoners are logically omniscient and to suppose that it is a new thing that they can learn is that some piece of evidence is logically related to some hypothesis. One can argue, in other words, that one thing Einstein learned *after* he formulated GR is that PMD is a logical consequence of it. But note that on this view as well, data used to tune a climate model cannot be used to confirm the model or any hypothesis

regarding the model, because climate scientists know what the relation between the data and model is as they do the tuning.

3) A third solution is simply to insist that all of the evidence that could possibly have been involved in model construction should be left out of the set B'. On this view (which is not, I would think for obvious reasons, a very popular one), it does turn out that data used to tune a climate model can be used to confirm the model or any hypothesis regarding the model.

Solutions 1 and 2 above correspond to what philosophers of science call "predictivism." Predictivism is the view that the amount of confirmation that some piece of evidence provides vis-à-vis some hypothesis can depend, in some cases, on whether the hypothesis predicted the evidence or accommodated it.

In *The Paradox of Predictivism*, Eric Barnes distinguishes several different versions of predictivism (Barnes 2008). According to "tempered weak predictivism," predictive success is not intrinsically superior to accommodation. But it *can* be a symptom of a property that the theory has that is hidden. That feature, in turn, can carry epistemic significance. According to strong predictivism, on the other hand, the history of the way a theory was put together can take central stage in the evaluation of the theory. The same evidence, according to strong predictivism, can provide different degrees of confirmation depending only on whether the theory's creator knew about that evidence at the time she constructed the theory. Strong predictivism is unappealing, according to Barnes, because it suggests that the degree of support afforded a theory depends on biographical facts about its creator. Weak predictivism, on the other hand, avoids this. It grants to the non-predictivist that what is ultimately responsible for a theory has to be a feature that is internal to the theory. If two theories are accorded different degrees of belief by us in the light of the same evidence, it can only be because one theory has feature F and the other doesn't. It can't be merely because theory 1 predicted the data and theory 2 only accommodated it. But tempered weak predictivists explain how predictive success can still play an epistemic role. Predictive success by theory 1 can be evidence that theory one has some feature F that we were unaware of.

John Worrall's famous coin-flipping example helps to make this idea clear (Worrall 2014). Imagine a coin is flipped 100 times. Scientist A predicts all 100 coin flips. Scientist B only predicts the 100th flip but her model can accommodate all the previous flips. Should we trust Scientist A more than B to predict the next 20 tosses? The strong predictivist says "absolutely." The tempered weak predictivist says "hold on a minute! Are there any other significant differences between A's model and B's model? If not, then I am just going to conclude that A got lucky. But if there might be differences between A's model and B's model that I am not well positioned to evaluate myself, then I might use the fact that A made good predictions as a guide to what 'good-making' features A's model has."

In the coin-flipping case, for example, the tempered weak predictivist might conclude from the fact that A has such impressive predictive successes, that A's model probably has some good-making features: maybe a good model of the tossing mechanism and how it produces flip outcomes. In other words, to a tempered weak predictivist, predictive success is not intrinsically epistemically significant, but it can be epistemically significant as a symptom of some other epistemically significant feature – especially if the bystanding observer is not well positioned to evaluate all of A's model's intrinsic strengths and weakness.

There are two features of climate models and their application that make them especially good candidates for falling into the category of hypotheses that the tempered weak predictivist acknowledges get a boost for predicting, rather than accommodating, data sets. The first one is noted by Frisch, appealing to earlier work of mine and Johannes Lehnard (Lenhard and Winsberg 2010), which we discussed in Chapter 5: climate models are, to some degree epistemically opaque to their users and developers. That is: the features that account for the successes and failures of climate models in simulating certain aspects of the climate system are not always fully under-stood and appreciated by modelers who use and make them. In Barnes' terms, this makes climate models especially strong candidates for being the kinds of entities that have hidden underlying features that are "good-making." If this is right (Frisch 2015, p. 190) argues in more detail than I do here that it is), then the concerns of even the tempered weak predictivist about when prediction vs. accommodation matters are relevant here, and there is an epistemic asymmetry between prediction and accommodation.

The second difference has to do with the difference between truth and model adequacy. What is typically being evaluated in the model evaluation process – the one that the climate scientists are discussing in the quotations above – is the model's adequacy for purpose, not its overall truth, or gen-eral fit with the world. Climate scientists usually already have a pretty good idea of which parts of their models are veridical and which parts are idealizations. What they are ignorant of when they do model evaluation, which they are keen to find out, is how much those idealizations impact the skill of the model for making predictions of various kinds.

As I pointed out earlier, all Frisch argues for regarding climate models is what he calls the "weaker claim" that "data not used in tuning can be more highly confirmatory of a model or provide a superior test in model evaluation than data used in tuning" (Frisch 2015). He does not endorse the "stronger claim" that "data used in tuning simply cannot be used in model evaluation or in the confirmation of models at all." But most climate modelers who have weighed in on this seem to endorse the stronger claim. What explains this? And who is right? I think the correct answer depends on paying closer attention to the fact that climate model evaluation is evaluation of *skill*, and not overall *fit*. Recall that we pointed out that the mathematical reasoning that Steele and Werndl employ is independent of whether we consider the prior and posterior probabilities to be probabilities of the model themselves, of the hypothesis that the model is "valid" or "confirmed" or whatever, or of the hypothesis that the model has some degree of skill.

When he is discussing Steele and Werndl's argument, Frisch is very careful to distinguish the question of the confirmation of a *model* from the question of the confirmation of *a hypothesis regarding model skill*. But I think he then goes on to fail to appreciate its significance. When we move from Steele and Werndl's maximally anti-predictivist framework to the framework of tempered weak predictivism, we might need to revisit the question of whether this distinction makes a difference. But Frisch does not do this.

Let's revisit the case of the coin-flip prediction model. Suppose that coin-flipping scientist Elizabeth is demonstrating the virtues of her model to Frank and Gertrude. Frank doesn't know any physics, but Gertrude knows all the fundamental physics that drive coin flips. She knows all the physics and modeling methodology that Elizabeth is making use of. (Let's suppose, just to keep things simple, that Elizabeth and Gertrude assign probability 1 to the fundamental physics they believe in.) The problem (for both Elizabeth and Gertrude) is that the physics is not entirely computationally tractable. But Elizabeth has come up with an idealized model that is partly based on the fundamental physics and partly based on a parameterization trick. Suppose further that her parameterization depends on details of the coin-flipping situation, and the only way she knows how to get her model to work with a particular coin-flipping situation is to tune it to the outcome of 100 trials. Frank and Gertrude both watch her flip the coin 100 times, tune the model, show that the tuned model can correctly retrodict the 100 past flips, and then successfully predict the next 100 flips.

Question: According to a tempered weak predictivist, did the fact that she could successfully retrodict the 100 flips confirm her model or are only the second 100 flips confirmatory?

Answer: It depends on if you are Frank or Gertrude. Depending on the details, if Frank is a tempered weak predictivist, he might very well say, "yes, the first 100 flips are confirmatory." Relative to his background knowledge, he might (again, depending on many details) think that the ability to retrodict is impressive. To him, the success might be a symptom of hidden good-making properties of the model (e.g. good grounding in physics) that he was unaware of. But Gertrude is not in this epistemic state. Let's suppose that she assigns probability one to the fundamental physics that guided the model, and probability zero to all the idealizations. She might smile knowingly at Frank and think to herself, "how quaint that to Frank, the fact that the model construction was guided by true physics is a hidden good-making feature. And how quaint that he uses Elizabeth's ability to retrodict as a guide to this." But she might have the relevant background knowledge to be in a position to be completely unimpressed by Elizabeth's "curve fitting." She thinks, "yes, of course Elizabeth can backfit a set of 100 flips. After all, her model has some veridical parts, and it has enough flexibility to fill in the wrong parts come what may." All she wants to know is: do the falsehoods that she knows are in Elizabeth's model offset in the right kind of way to give her model skill? Or do they only provide it with the flexibility to backfit (in a way that she never doubted)? If that is the epistemic state she is in, then she will adopt Frisch's "stronger claim" for herself, all the while conceding that for Frank, only Frisch's "weaker claim" is true.

When Schmidt and Sherwood say "characteristics (or metrics) that are used to explicitly tune a model or its inputs should not also be used to evaluat[e] the model," (2015, p. 159) I think we need to read them as speaking from the epistemic vantage of point of someone like Gertrude. Here is one way to make the point: the fact that a climate model can be tuned to some existing data set tells you some things about the goodness of the model. After all, a model with completely wrong physics in it, or one that was structurally defective in fundamental ways, would not be tunable. So, if you had poor background knowledge about a climate model, then the fact that the model could be tuned to an existing data set might boost your degree of belief in some hypotheses regarding some of the features of the model. As lay observers, this might very well be the state we are in. This is not, however, the epistemic state that climate modelers are in when they test their models for skill, after having already tuned them. At that point in the process, they are almost exclusively concerned with the degree to which features they are unaware of are helping or hindering model skill with respect to some variable or set of variables. (The degree, to use Barnes' language, to which the models have hidden good-making features with respect to model skill of some particular kind.) If so, then Frisch might be right that, from a particular epistemic vantage point (indeed the one that we lay observers have), the fact that a climate model can successfully be tuned boosts its credibility in various ways (albeit not as strongly as novel prediction). From that vantage point, only the "weaker claim" might be true. But from the epistemic point of view of a modeler who has built the model and is only interested in model skill relative to competitors, the stronger claim might apply. That is, such an epistemic agent might be right that no additional confirmation occurs as a result of being able to tune the model relative to where she was before.

## 10.5 Conclusion

Evaluating model skill is the central topic in the epistemology of climate models. This is not a matter of determining the "correctness" or "fit" of climate models, but rather their suitability for various prediction and projection tasks. While the "gold standard" for establishing model skill is the "verification and validation" paradigm, that paradigm is rarely helpful in climate science, because it does not apply well to modeling projects in which there is substantial back and forth between model evaluation and model construction (as illustrated in Figure 10.2). Rather than following an articulable set of normative rules, establishing the credibility of model-based knowledge in climate science is local and detailed, even though this claim might not sit well with those who might hope for an epistemology that can be reconstructed or rationalized in terms of generally recognized, ahistorical evidential norms.

This does not mean, however, that nothing general, global or ahistorical can be said about the epistemology of climate models. We can, for example, ask questions like the one we discussed in section 10.4: How much reliability can we infer from the ability of a simulation to reproduce data against which it was tuned? The answer we argued for was that the ability to predict novel data is always a stronger test of model skill than being able to re-accommodate data that were used in the process of tuning a model. But we also argued that the question of whether reaccommodation is *at all* indicative of model skill might depend on one's epistemic position.

## Suggestions for Further Reading

Goodwin, William M. 2015. "Global Climate Modeling as Applied Science," *Journal for General Philosophy of Science*. An incisive paper that argues for a "relative strategy" for securing the epistemic reliability of climate models. The strategy is to try to show that they are epistemically on a par with other scientific techniques that have not been called into question.

Mauritsen, Thorsten et al. 2012. "Tuning the Climate of a Global Model," *Journal of Advances in Modeling Earth Systems*. An indispensable resource for understanding the practice of tuning in climate modeling.

Barnes, Eric Christian. 2008. *The Paradox of Predictivism*, Cambridge University Press. An excellent monograph on the topic of predictivism – the thesis that predicting evidence is more confirmatory than accommodating it.

Schmidt, Gavin A. and Steven Sherwood. 2015. "A Practical Philosophy of Complex Climate Modelling," *European Journal for Philosophy of Science*. A very nice *philosophy* paper by two climate scientists. Offers the climate scientist's take on two of the philosophical topics we have discussed in the book: values in science and the issue of "double counting" old evidence.

Steele, Katie and Charlotte Werndl. 2013. "Climate Models, Calibration, and Confirmation," *The British Journal for the Philosophy of Science*. Argues that climate scientists are confused about the issue of double counting old evidence by appealing to the Bayesian framework of confirmation theory.

Frisch, Mathias. 2015. "Predictivism and Old Evidence: A Critical Look at Climate Model Tuning," *European Journal for Philosophy of Science*. A response to Steele and Werndl, above. Argues that the Bayesian framework itself is fundamentally silent on the question of how to deal with old evidence and that there are features of climate models that make using old evidence for confirmation especially problematic.

# *Robustness*

## 11.1 Introduction

Recall the "Guidance Note for Lead Authors of the IPCC Fifth Assessment Report on Consistent Treatment of Uncertainties" that we discussed in Chapter 7. The document counsels that assessments of *confidence* regarding climate hypotheses ought to reflect "the type, *amount*, quality, and *consistency* of evidence (e.g., mechanistic understanding, theory, data, models, expert judgment) and the *degree of agreement*." (Mastrandrea et al. 2010, p. 1). Indeed, it has been widely noted that Pirtle, Meyer, and Hamilton found 118 articles in which the authors used the *agreement* between a wide variety of climate models to encourage *confidence* in the results of the models(Pirtle, Meyer and Hamilton 2010, p. 3).

In this chapter, I want to discuss the features of this recommendation that I have emphasized in the above quotation. These facts describing scientific practice have remained something of a puzzle to philosophers of science. Why, exactly, do we have more confidence (or even a higher degree of belief) in a hypothesis when there is strong "agreement" amidst "more" and "consistent" evidence in favor of the hypothesis? Perhaps more puzzlingly, what exactly counts as having "more" evidence?

Imagine that you have read in the *Sioux City Journal* that O.J. Simpson's glove was found at the house of his ex-wife Nicole Brown. This evidence gives you some reason for believing that Simpson killed his wife. Suppose I want to add to your degree of belief in the hypothesis by giving you more evidence. What would count as "more"? One cannot simply count the number of tokens of evidence. One way you could have more tokens of evidence for the hypothesis, after all, is that you could have in your possession more than one copy of the issue of the *Sioux City Journal*, each of which reports the same fact: that the glove was found at the scene. Surely by the fourth copy, however, having a fifth, sixth, or seventh copy of the same paper reporting the same fact ceases to relevantly count as "more evidence" of O.J.'s guilt. Perhaps reading similar reports in *The New York Times*, *The Washington Post*, and the *Quad-Cities Times* counts as more evidence. But again, after reading a certain number of different publications that report the same fact about the glove, we begin to doubt that adding more news outlets counts as piling on "more evidence." Eventually, in order to count as having seen "more evidence" of O.J.'s guilt, we would like to see reports of hair fibers, shoe prints, blood samples, and the like. And the conspiracy-theory-minded among us might want to see more than media reports of these facts. She might want to see for herself that they were in fact presented in court. She might even want to see evidence that the physical samples presented in court had followed the proper chain of custody, that the police officers who handled the evidence were not racists, etc. All of this is much better than having five, nine, or 22 copies of the *Sioux City Journal*. These are fairly banal observations about what counts and does not count as "more evidence" of the guilt of an accused person. But what, exactly, makes these observations true?

When it comes to issues surrounding the question of what counts as "more evidence" of a climate hypothesis, a fair amount of debate has emerged in the philosophical literature. One reason that the issue is less easy to settle in the case of climate hypotheses is this: it is hard to know, in the case of collections of climate models, how much they are like copies of the same newspaper, or how much they are like gloves, hair samples, and shoe prints. In other words, if several different climate models all show that we should expect the ice caps to melt at such-and-such a rate, is that more like reading the same fact in several different copies of the same newspaper? Or is it more like accumulating genuinely distinct pieces of evidence in a criminal trial? In the case of climate hypotheses, it would appear, the question is more complicated, and debate has emerged. It would be nice to have a *systematic* account of why, and under what conditions, more tokens of evidence count as more evidence that ought to be more persuasive – this would help us to understand what we learn from the agreement of multiple models and the like in climate science.

In the philosophy of science, debate surrounding these topics has mostly taken the form of attempting to connect these issues in climate science and climate modeling with philosophical accounts of what has come to be known as "*robustness analysis*." In this chapter we will therefore take a look at different accounts in the philosophical literature of robustness analysis (hereafter "RA"), and at attempts to connect them to climate science, and try to see if it can help us get a handle on the following question: when and why does a diversity of evidence provide stronger support for climate models and/or for climate hypotheses?

## 11.2 Robustness Analysis – A Focus on Models

Discussion of RA entered the philosophical literature via a contribution from the biologist Richard Levins. Levins was concerned with how we reason from multiple models. He claimed that RA could tell us "whether a result [we derive from a model] depends on the essentials of the model or on the details of the simplifying assumptions." (Levins 1966, p. 423). If we study the results of a number of different models, and each of them, "despite their different assumptions, lead to similar results, we have what we can call a robust theorem that is relatively free of the details of the model. Hence, our truth is the intersection of independent lies"(ibid.).

Michael Weisberg summarized Levins' take on RA as follows:

> When confronted with multiple models of the same phenomenon, the theorist needs a way to separate reliable predictions from artifacts of the assumptions made by the different models. Thus the theorist examines multiple models, preferably ones which make qualitatively different assumptions about the phenomenon, and looks for a common prediction among these models. This common prediction is the "truth at the intersection of independent lies," or a robust theorem. (Weisberg 2006, p. 732)

According to Weisberg, RA is only needed in the absence of "comprehensive theory" because "when dealing with simple, physical systems, one of the fundamental physical theories could be used to determine how much distortion was introduced by each idealization. Such theories have the resources to estimate the effect of various idealizations, providing guidance about what must be included when particular degrees of accuracy and precision are required" (Weisberg 2006, p. 731).

It is not clear, however, what work the word "simple" is doing in the above, since students of physical systems (like the climate) do indeed have at their disposal fundamental physical theories, yet they still often need help in determining how much distortion each idealization adds. It is not obvious, moreover, what role complexity plays in preventing them from being used to determine how much distortion is introduced by each idealization.

Fundamental theory, on its own, cannot always tell us (perhaps it doesn't even often tell us) much about which idealizations will introduce distortions and which won't. This usually has to be checked empirically, or by some other means. As a result, even though we are concerned with a system (the climate) in which there is something akin to "comprehensive theory," the literature on the philosophy of climate science has been concerned with the extent to which RA could be used to support the predictions that follow from an ensemble of models. As such, even though we are in the domain of comprehensive theory, we should still take seriously a famous criticism of Levins' proposal that came from Steven Orzack and Eliot Sober (Orzack and Sober 1993).

Orzack and Sober worried that Levins was trying to provide a means of confirming hypotheses in the absence of empirical evidence. And sometimes it sounded like he was. In any case, as it stood, Levins' account of RA was vague. In order to criticize the account, they reconstructed it as follows: Let us call the various models that the employer of RA is relying on $M_1...M_n$, and we will call the set containing all of these models $\mathbf{M}$. And let us suppose that the RA in question is being used to support some hypothesis H. If H follows from, or can be calculated using, $M_i$, then we will say that $M_i$ "detects" H.

Even though they were setting it up for defeat, let's call the following the Orzack and Sober account of RA.

Levins' RA principle: if all the members of $\mathbf{M}$ detect H, then H is a "robust theorem" and H is likely to be true.

Orzack and Sober argued that anyone who wants to employ this principle faces the following dilemma: unless the employer of RA knows that at least one member of $\mathbf{M}$ is the "true model," then the fact that all of the members of $\mathbf{M}$ detect R is no reason at all to believe R. But if she already knows this, then it is unclear what role the rest of the set is playing. We should keep this criticism in mind as it reappears in the debate about RA in climate science.

## 11.3  Robustness Analysis in Climate Science: Lloyd and Parker

When it comes to the question of the applicability of RA in climate science, papers by Elisabeth Lloyd (Lloyd 2009, 2010, 2015) and Wendy Parker (Parker 2011) have been at the center of the debate. An over-simplified gloss on the debate would be to say that Lloyd has argued that RA can be used to support climate models and climate hypotheses, and Parker has demurred. A somewhat deeper analysis would reveal that Parker has been mostly keen to argue that the fact that a hypothesis about the climate is supported by a given *ensemble of models* is not, by itself, sufficient grounds to accept the hypothesis. Lloyd, on the other hand, has been keen to argue that RA can be part of a more comprehensive story about how a variety of evidence *and* the agreement of models can provide rational grounds to accept the hypothesis. Part of the work of bringing these views into contact with each other seeing how the different things they say are in fact compatible, therefore, will involve getting a grip on two different questions: 1) How much work can RA do in understanding whether diverse models support climate hypotheses? 2) How much work can RA do in understanding how diverse models *and other sources of evidence* work together to support climate hypothesis?

In the sequel, we will agree with Parker on the first question up to a point, but we will argue that she has overlooked ways in which model ensembles can be used to create robust evidence. We will come to agree with Lloyd on the second question, but our account will differ considerably in the details. One key difference with Lloyd and Parker, as we will see, is that we will abandon the idea, as we find it in Levins and Weisberg, that RA is exclusively about analyzing the phenomenon of *model agreement*. Still, it will be very helpful to briefly review the exchange between Lloyd and Parker.

Lloyd first introduced the idea of RA being applied to climate science in a paper (2009) in which she made the argument that climate models and the findings we derive from them are supported by a variety of empirical means beyond mere goodness of fit. In addition to fit, she argued, climate models receive additional support from the fact that they are supported *by a variety of evidence, and* that their results are supported by the fact that many of these results are *robust under a whole class of models*. She noted, for example, that the fact that all global climate models simulate global warming of 0.5–0.7°C for the twentieth century – the fact that this result is robust under all the models – provides support for the claim that we can trust them regarding other predictions. Lloyd, in other words, was pointing to a synergy of two complementary kinds of evidential support: the support that a variety of empirical evidence gives to our models, and the support that our models give to their predictions in virtue of the fact that multiple models agree regarding certain predictions. Considerations of robustness, nevertheless, were restricted to the second kind of support in Lloyd's account. In our approach, we will agree with Lloyd that the most interesting question is how the synergy of evidence works – but we will apply RA more broadly in our answer – not limiting it to the second kind of support.

Though Lloyd and Parker were addressing slightly different issues, it is still tempting to treat them as the robustness booster and the robustness skeptic, respectively. As a robustness skeptic, Parker (2011) offered a variety of considerations designed to motivate skepticism concerning the relevance of RA to model ensembles. Her work begins by considering an analysis of RA that closely matches the one we are calling "the Levins account" (as reconstructed by Orzack and Sober). Recall that, in that account, in order for RA to support a result that follows from all of the members of **M**, we have to have confidence that at least one member model of **M** is the "true model." As Parker points out, this will definitely not work in climate science, and it is unlikely to work in much science at all, because the idea of a "true model" is inapplicable to most science. "Insofar as a scientific model can be identified with a complex hypothesis about the workings of a target system, there is usually good reason to believe that such a hypothesis is (strictly) false since most scientific models are known from the outset to involve idealizations, simplifications, or outright fictions" (2011, p. 584).

This is exactly right – both about climate models and about models in science in general. Back in Chapter 3, we agreed with Parker that climate models are best discussed in terms of their adequacy for purpose, rather than their truth, or overall fit. At the very least, therefore, we are going to have to replace the notion of *truth* in this picture of RA with the idea of *adequacy for purpose*. Parker has a helpful suggestion that we can reconstruct as follows:

Suppose that we are interested in some specific hypothesis about the climate, H. H might be the hypothesis that the addition of a heat-trapping gas to a planet like the earth will warm the tropical troposphere. Once again, we will say that a model $M_i$ detects H just in case H is a result of $M_i$, – that is that H can be derived, calculated, or inferred from $M_i$. And we will say that $M_i$ is "capable of correctly detecting" H just in case H is adequate for the purpose of determining the truth of H to a particular desired degree of accuracy. Thus we might say that a climate model is "capable of correctly detecting the warming of the tropical troposphere under carbon forcing" if we have in mind that the model is adequate for the purpose of determining how much the tropical troposphere will warm under a forcing to within a specific degree of accuracy. We can now alter the Levins requirement for successful RA: in order for RA to demonstrate H, we have to show that is likely[1] that at least one member of **M** is correctly detecting H. Here, what we mean by some M correctly detecting H is twofold.

---

[1] In addition to weakening the demand that M be true, to the claim that it be capable of H, we are also weakening the (probably unreasonable) demand that RA be able to demonstrate that M has this

1) M is capable of veridically detecting H to the degree of accuracy with which H is specified.
2) M actually detects H.

Unfortunately, this reconstruction of RA is still not going to work well for climate science. Parker explains that there are two reasons for pessimism here. The first is that, for most of our interesting climate hypotheses, we do not, in fact, have particularly strong reasons for thinking that any one single climate model in the set of available models is especially capable of detection with respect to any one of those hypotheses. In effect, this is the reason *in the first place* that we might want more models.

The second reason for pessimism runs deeper, and highlights why Orzack and Sober's analysis of Levins' reasoning is meant to be so undermining: nothing about what we have said so far gives us any reason for thinking that the *number* of models we have is making it any more likely that *one of them* is veridical regarding H. In other words, on their reconstruction, we have a dilemma: either we already trust one model, or we don't. If we do, then the variety of models is irrelevant. And if we don't, then RA is not going to help us establish the likely truth of H. What is entirely missing from the analysis so far, in other words, is anything like a compelling story about how, as the number of models in M goes up, the likelihood that one of them is capable of correctly detecting H goes up too.

Here's a possible analogy. Imagine you are at a party with hundreds of guests, and you want to know what time the last streetcar will take you home. You want to ask one of your fellow guests, but you are worried that he or she might be misinformed about the correct time. You ask a few dozen guests and they all say 11pm. One reason you might trust this report might be if you had reason to believe that at least 30 percent of the guests had carefully read the schedule; and you might have reason to think that your sample of guests was random. This vignette is meant to be analogous to the idea that the models are in some way sampled from the space of possible models, one of which must indicate correctly, since there must be a possible model that indicates correctly. Setting aside the question of whether that last clause is even true, Parker argues quite convincingly that it is a non-starter to entertain the idea that ensemble methods in climate

property to the demand that it be likely that M has this property. (Perhaps to the demand that it raise that likelihood above some critical threshold for acceptance – one perhaps dictated by context.)

science sample randomly from the space of possibility. She quotes climate scientists who point out that our climate model ensembles are "ensembles of opportunity" and not anything like random samples. In the end, Parker concluded that "while there are conditions under which robust predictive modeling results have special epistemic significance, scientists are not in a position to argue that those conditions hold in the context of present-day climate modeling" (Parker 2011, p. 597).

---

### Box 11.1 Knutti on Process Understanding

Physicist and climate scientist Reto Knutti has authored a paper in which he articulates his "philosophical perspective" on numerical modeling (Knutti 2018). In it he argues that neither "model fit" (showing agreement between model predictions and observed data) nor "robustness" (which he interprets, following Parker, as agreement in prediction between the models in our "ensemble of opportunity") can give us confidence in the skill of our models to make accurate predictions outside the range of forcings for which we have data. He argues that acquiring genuine confidence that our climate models have these kinds of skills requires us to have what he calls "process understanding."

To understand what Knutti means by process understanding, we need to understand that although climate models are driven by a mixture of basic physics, and models of other underlying components of the climate system, there are also various features and variables of the climate that are emergent. The emergent features and variables arise out of the underlying components rather than being given in them. Process understanding comprises having knowledge of the relevant quantitative relationships and interactions between different emergent components of the system. But it also comprises having well-justified beliefs about how those relationships and interactions will or won't be preserved as we move into time periods or regimes outside those for which we have data that we can use for evaluation. And it also involves having confidence that we have not neglected any other important relationships and interactions. Only once we have all of this, he claims, can we reliably infer from the fact that a model is adequate for predicting some domain of behavior for which we *have data*, to the claim that the model will be adequate in domains for which we don't.

Even though Knutti's claim is intuitively plausible, it would be nice to have a better understanding of why it's true, or at least why we should think it's true. In the next chapter, I'll argue that, *pace* Knutti and his agreement with Parker, it's only on a proper understanding of *robustness* that we can understand why process understanding can *confidently* project confidence in model skill into untested domains.

## 11.4 Robustness Analysis – Beyond Model Agreement

As I hinted at above, one striking thing about the exchange between Parker and Lloyd on robustness in climate science is the degree to which the targets of their discussions seem to differ. Lloyd seems primarily interested in understanding how a whole spectrum of evidence in climate science supports its conclusion, and Parker is primarily interested in the phenomenon of "when models agree." These are of course not the same question. But a commitment to a conception of RA on which it is only applicable to the phenomenon of *model agreement* goes some way to obscuring this difference.

This is an unfortunate consequence of the way the literature on RA got off the ground. The original piece by Levins, the criticism of it by Orzack and Sober, and its resuscitation by Weisberg, all concerned the idea that we could find "truth at the intersection of independent lies." The "lies" Levins was interested in were idealized *models*. Similarly, Parker's arguments against the usefulness of RA in climate science were presented in a paper titled "When Climate Models Agree," and as such was concerned only with the question of whether support for climate hypotheses could come from the fact that they follow from a *variety of models*.

But this is not the only way in which models, simulations, data and observations combine in order to support climate hypotheses. Consider the hypothesis that equilibrium climate sensitivity (ECS) lies between 1.5°C and 6.0°C. This hypothesis is supported by the fact that all of our best climate models produce outcomes within this range, and, to be sure, those models are in turn supported by a variety of evidence. But the hypothesis itself is also supported, directly, by a variety of paleoclimate data, instrument records, theoretical reasoning, and the like.

Luckily, there has been a fair bit of discussion, in the philosophical literature, on the applicability of RA to laboratory and experimental measurement, observations, and experiments. While most have argued, (cf. e.g. Woodward 2006, Calcott 2011) that these represent entirely different sorts of activities, Jonah Schupbach (2016) has recently argued persuasively that one single conception of RA is an equally useful tool for analyzing experiments, observations, inferences and derivations from models (and simulations), and the like. This is an incredibly useful result for us; most of our best climate hypotheses are supported by a *number of epistemic activities*: models, simulations, weather instruments, proxy data, theoretical calculations, etc. As we will see, if we really want to get a handle on how diversity of evidence works in climate science, it would be best if we could come up with an account of RA that was ambivalent in precisely this regards – one that could equally well treat sets of models, sets of experiments, sets of observations, or most importantly, heterogeneous sets of all of the above. Schupbach's account is so useful for us precisely because it works in just that way. But it is equally useful for understanding how model robustness arguments could actually function in climate science. The rest of what follows will lean heavily on his analysis.

*Pace* Parker's take on robustness in "When Climate Models Agree," (Parker 2011) Lloyd's work on robustness in climate science has always tried to take into account the importance of the variety of types of evidence. This is especially true in her 2015 work. In this respect, some of our conclusions will be similar to hers. We part ways with her, however, by leaving behind the idea, which one finds in Levins, Weisberg, and other recent philosophical literature[2] on robustness, that RA is restricted to analyzing the situation in which a diverse set of *models* result in the same finding. Rather than trying to show how RA is *part of* a complex epistemic landscape in which other sources of evidence also play a role, our aim will be to show how RA can offer a comprehensive picture of that complex landscape – of how all these sources of evidence work together.

## 11.5  Sufficient Diversity of Evidence – Cumulative Epistemic Power

Regardless of whether we are interested in model robustness alone, or in robustness as a picture of a wider range of evidence, we need to find a better link between increasing quantities of evidence and reliability than those that are canvassed by Parker. As the O.J. case is meant to show, something about the *quality of diversity* of the evidence is going to have to be mustered if the quantity of it is going to do the desired work. Weisberg puts the point in terms of needing a "sufficiently diverse set of models." (Weisberg 2006, p. 732) But what kind of diversity, exactly, are we looking for? And for what purpose? For the moment, we will follow most of the literature in focusing on models, and expand our search later on to include other sources of evidence. Having said that: what property of a set of models makes it "sufficiently diverse" such that its size will correlate with its reliability?

---

[2] The exceptions are Wimsatt (2011), Woodward (2006) and Calcott (2011) who talk about RA in the context of experimental reasoning and other contexts, but they take them to be different forms of RA.

There are two ways we could try to spell out the relevant notion of the aim of diversity. Most of the literature has focused on this way of spelling it out:

1) Recall our definition of detection: some model $M_i$ detects a hypothesis H just in case H can be derived, calculated, or inferred from $M_i$. We could characterize the aim of "sufficient diversity" in the following way: we want it to be the case that if a set **M** of models is "sufficiently diverse," and all the members of **M** detect H, then we could rationally conclude that H is true.

But trying to achieve a definition of diversity that meets that standard seems optimistic. Most recent philosophy of science has tried to acknowledge that science at best offers grounds for increasing one's degree of belief in a hypothesis, and that standards of hypothesis *acceptance* generally depend on context and other factors. In line with that thinking, I would argue that we should instead look for a notion of diversity that emphasizes *confirmation* (raising our credences) rather than *acceptance*.

2) Alternatively, instead of defining a notion of "sufficient diversity," we might also try to identify a notion of "RA diversity". The idea would be to define the notion such that if a set of models **M** that indicate H is RA diverse, then each time we add one more model to the set our confidence in H goes up.

The difference is substantial.

## 11.6 Conclusion: RA Diversity

Continuing to restrict ourselves, for the moment, to the domain of *model agreement*, let's try to make notion #2 a bit more precise. In line with the idea that we get more confidence each time we add one more model, we will call the notion "cumulative epistemic power":

Let $D_k$ be the proposition that model $M_k$ detects H. We will say that a set **M** containing models $M_1 \ldots M_n$ has cumulative epistemic power (CEP) to detect H just in case it has the right property for making it rational to conclude that as long as, when i<n

$$\Pr(H|D_1..D_i) < \Pr(H|D_1 \ldots D_{i+1})$$

What we are looking for, then, is a kind of diversity that a set of models might have such that possession of it would give us grounds to believe that that set has CEP. Ideally, of course, this would be a kind of diversity that we can actually find in climate science. If we can find it, we will call it "RA diversity." We should distinguish the notion of RA diversity from "sufficient diversity." Crudely, RA diversity doesn't speak to whether a set of detection methods is large enough. It only speaks to the set gets better as it gets larger. But if we have a notion of RA diversity at our disposal, then we would know exactly when to expect each successively new model to augment our degree of belief in H, and we can let the particular context of an epistemic situation settle the question of how much RA-diverse evidence we need to collect before we should accept H.

So much for model diversity. Can we expand that notion beyond the domain of detection by *models*? Our interest is in possibly heterogeneous sets of *any sort of procedure at all*, regardless of whether the set contains models, observations, experiments, simulations, etc., or any mix thereof. To do this, we will need to modify our definitions of detection and of CEP.

We will say that a procedure P (instead of specifically a model M) detects H just in case performing P points at the truth of H. This could be because we derive H, or we observe H, or we calculate H from what we observe, or we infer H from our model, etc. Now let $D_k$ be the proposition that $P_k$ detects H. We will say that a set **P** containing procedures $P_1...P_n$ has cumulative epistemic power (CEP) if it has the right property for making it rational to conclude that as long as i<n

$$Pr(H|D_1..D_i)<Pr(H|D_1...D_{i+1})$$

Luckily, I believe one such account of RA diversity – one that is independent of what kinds of procedures are involved, and that meets all of the above desiderata – is available, and can rationally ground a claim of CEP as we will see in the next chapter.

## Suggestions for Further Reading

Levins, Richard. 1966. "The Strategy of Model Building in Population Biology," *American Scientist*. The original paper where the idea of model robustness was introduced.

Orzack, Steven Hecht and Elliott Sober. 1993. "A Critical Assessment of Levins's the Strategy of Model Building in Population Biology (1966)," *The Quarterly Review of Biology*. The paper in which Orzack and Sober argue that Levins' idea is unworkable.

Lloyd, Elisabeth A. 2009. "I – Elisabeth A. Lloyd: Varieties of Support and Confirmation of Climate Models," in *Aristotelian Society Supplementary Volume*. Lloyd's first argument in favor of using the philosophical literature on robustness analysis to understand how climate scientists use model ensembles to support climate hypothesis.

Parker, Wendy. 2011. "When Climate Models Agree: The Significance of Robust Model Predictions," *Philosophy of Science*. Parker here argues against the confirmatory value of model robustness in climate science. The paper goes through a variety of ways of understanding robustness arguments and argues that none of them are applicable in climate science.

Lloyd, Elisabeth. 2015. "Model Robustness as a Confirmatory Virtue: The Case of Climate Science," *Studies in History and Philosophy of Science Part A*. Lloyd revisits the question of the confirmatory value of robustness in climate science.

Schupbach, Jonah N. 2016. "Robustness Analysis as Explanatory Reasoning," *The British Journal for the Philosophy of Science*. A clear and cogent argument for understanding robustness (both of the model and experimental variety) in the manner that I have defended in this chapter. A must-read for anyone interested in robustness analysis.

## 12

# *Diversity*

## 12.1 Introduction

We left the last chapter wondering what property a set of detection methods has to have in order to count as diverse for the purpose of Robustness Analysis (RA). Unfortunately, the literature on this topic has been plagued by an assumption that has limited progress on this question. In keeping with Levins' intuition that RA is about finding "truth at the intersection of *independent* lies," the literature has mostly assumed that what we need for a set of detection models to be RA-diverse is some kind of *probabilistic independence*. Schupbach (2016), however, has argued persuasively that simple probabilistic notions will not work. To see why, it will be useful to look at two such proposals.

### *1) Unconditional Probabilistic Independence*

Levins characterized the results of RA as "truth at the intersection of independent lies." Many have taken this to indicate that the kind of diversity we are looking for in RA is a kind of probabilistic independence. The simplest such proposal we could consider is one that defines two means of hypothesis detection as diverse with respect to RA just in case they are straightforwardly probabilistically independent of each other. In other words, suppose we have a set $\mathbf{P}$ of detection procedures (these could be models, observations, experiments, etc.) containing procedures $P_1..P_n$. And suppose H is a hypothesis we are interested in. Again, let $D_k$ be the proposition that H is detected by $P_k$.

On the proposal we are considering here, the set $\mathbf{P}$ is RA diverse just in case for any i, j, the probability of $D_i$ is independent of $D_j$. In other words, just in case:

$$Pr(Di)=Pr(Di|Dj) \text{ and } Pr(Dj)=Pr(Dj|Di).$$

This is the notion of independence that Orzack and Sober considered when they evaluated Levins' conception of RA. It is also a proposal that Parker considers in her argument against the applicability of RA to climate science. It is not very surprising that both Parker, and Orzack and Sober, found this purported notion of RA diversity inadequate. If we think about what the condition entails, then we can see that it will rarely, if ever, be met under the sorts of circumstances where we might expect RA to work. Take the O.J. case. The fact that we have read in the *Sioux City Journal* that O.J.'s glove was found at the crime scene obviously raises the probability that we will read the same thing in *The New York Times*. But insofar as we think it is at all indicative of O.J.'s guilt, it also raises the probability that we will have found his shoe prints outside the window. Similarly, in so far as two climate models share any basic properties at all (and we better hope that they do!), then the fact that one model detects some hypothesis will certainly raise the probability that the second one will. In sum, then, while unconditional probabilistic independence between the various Ps would certainly be good reason to believe H, this is of little help, since we do not expect that this condition will ever obtain – even in cases where we might intuitively think that the various Ps are RA diverse.

## 2) Independent Probability of Failure

In his 1994 work, William Wimsatt offered another notion of probabilistic independence as a criterion of RA diversity. His suggestion was that a set of detection procedures **P** should count as RA diverse, if and only if for any two members of **P**, Pi and Pj, the probability that Pi would fail (that is, that both Pi and ~H would obtain) is independent of the probability that Pj would fail. In other words:

$$\Pr(Pi|\sim H) = \Pr(Pi|Pj\&\sim H) \text{ and } \Pr(Pj|\sim H) = \Pr(Pj|Pi\&\sim H).$$

Let's call this criterion *independence of probability of failure*, or IPF. The idea here is that if we learn that one method of detection has misled us with respect to H, this should have, if the two methods are RA diverse, no effect on the probability that the second method will mislead us. It is easy to see the appeal of this proposal – it is certainly more plausible than unconditional probabilistic independence. If I happen to know that O.J. did not kill his wife, then it is appealing to think that, when I find out that the *Sioux City Journal* is reporting that his glove was at the scene, this should not raise the probability that his shoe prints will be outside Nicole Brown's window. Similarly, if a simulation model shows me that global surface temperature goes up by 2°C when I double carbon dioxide, *but I happen to know that climate sensitivity is zero*, then learning the first fact, it would seem, should not raise my probability that the paleoclimate record will show a correlation between $CO_2$ and temperature.

The problem is that while IPF is a powerful indicator of CEP, it is still too strong. It seems as if one can have CEP without IPF – and indeed we often do. Consider the O.J. case again: even if we assume that O.J. is innocent, whether or not finding his glove at the scene will raise the probability of finding his shoe print at the scene will depend upon the range of possible reasons why the glove could fail as a detector of O.J.'s guilt. Consider this possibility: maybe the glove was at the scene despite O.J.s innocence because someone was trying to frame him. Insofar as that is a possible explanation of the glove evidence failing as a detection of O.J.'s guilt, then it will not be independent of the shoe evidence. In general, insofar as two methods of detection of H *could be misleading for the same reason*, they will not exhibit independence of probability of failure. But they might still be RA diverse. For example, insofar as two pieces of evidence of O.J.'s guilt would both mislead on a single possible story of framing, we would not expect them to exhibit IPF. But that is no reason to disqualify them as being RA diverse, since they do seem to offer CEP, as the O.J. example indicates.

Or consider a hypothesis about the paleoclimate – perhaps something about the global temperature during a particular period – that is supported by a variety of climate data proxies – tree rings, ice cores, and coral samples. In general we would be inclined to regard the three samples as RA diverse with regard to a temperature record since they each provide, in some intuitive sense, independent evidence of the temperature, and they do seem to provide CEP. We would be inclined to think this precisely because, in general, something which caused the tree rings to be misleading *probably* wouldn't make the ice cores misleading. But the key word in the last sentence was the italicized "probably." It *could* be that the tree-ring data are misleading because a group of conspirators are trying to make the world believe something false about the paleoclimate, and have broken into climate labs to tamper with all three of these types of samples. The probability of the tree rings misdetecting H, therefore, is not probabilistically independent of the probability of the ice cores misdetecting. Of course the common confound needn't be so outlandish. It's not hard to imagine a single natural confound that would systematically skew all three of these sources of data – a volcanic eruption or a large meteor; the details needn't matter. Any time there is a *possible* confound that would skew two sources in the same direction, we would get a failure of IPF. And yet we do not want to conclude that the sources are not RA diverse merely for that reason.

### 3) Other Probabilistic Notions of Independence

There are various other ways of specifying the relevant notion of probabilistic independence that are discussed in the literature. At the end of the day, however, I think we should follow Schupbach in believing that they all suffer from the same problem: looking carefully at examples of plausibly RA-diverse methods of detection – whether they come from modeling, experiment, or ordinary reasoning like the O.J. case – reveals that intuitively RA-diverse methods fundamentally lack any interesting kind of probabilistic independence precisely because pairs of them often share features that make the probability of detection mutually dependent, conditional either on truth or falsity of the hypothesis. That they are probabilistically dependent conditional on the hypothesis is true is obvious. But they are also often probabilistically dependent conditional on the hypothesis being false because even when the hypothesis is false, there is usually a scenario of falsehood that will make any pair of RA-diverse methods likely to detect the hypothesis – as is exemplified by the example of someone having tried to frame O.J. for killing his ex-wife.

## 12.2  Explanatory Robustness

If we think clearly about what various RA-diverse methods of detection seem designed to do, it becomes hard not to notice the role that *competing hypotheses* play in motivating RA methods. This is the heart of Schupbach's analysis. Consider the case of consulting *The New York Times* after we read in the *Sioux City Journal* that O.J. Simpson's glove was found at the house of his ex-wife Nicole Brown. One reason that we might want to do this is that there are at least two competing hypotheses that might *explain* the existence of the above sort of report in the *Sioux City Journal*. One is O.J.'s guilt, but another is sloppy reporting on the part of the *Sioux City Journal*. Yet another might be that the editor of the paper has a personal vendetta against O.J. However, once we find a confirming report in *The New York Times*, sloppy reporting or a vendetta – specifically on the part of the *Sioux City Journal* – no longer seems like a plausible explanation. Still, we might worry about some systematic feature of the American press that would cause all news outlets to consistently misreport this kind of find. To eliminate this possible explanation, we might go down to the courthouse and see for ourselves that the prosecution did in fact present eyewitness police testimony that the glove was found at the scene. That would be an "RA-diverse" detection of O.J.'s guilt because, relative to the set we have already considered, it rules out the previously viable explanation that the press was systematically mistaken. Still, we might worry that the police are lying about this fact out of racism. That's an explanation not yet ruled out. To alleviate that concern, we might want to perform an investigation that would reveal whether an appropriate chain of custody was followed with the glove. Having performed all of these tests, we move on to considering possibilities that would explain the presence of the glove at the scene that does not involve O.J.'s guilt. Maybe it was not really his glove, and so we make him try it on. Or maybe it was his glove, but it somehow ended up at Nicole Brown's house because he loaned it to a friend. But this alternative explanation becomes less plausible when we find his shoe prints outside her window. And so on for each new piece of evidence. Each RA-diverse method of detection is meant to rule out a competing explanation (competing with hypothesis of O.J.'s guilt) of the results of the previous set of detection methods.

Does this work with climate hypotheses? It certainly does for paleo-climate hypotheses supported by proxy data. The whole point of collecting multiple sources of proxy data is to rule out the possibility of confounds that might distort one type of data. Suppose the tree rings we find that correspond to some range of years are narrower than for other years. One explanation for this could be a cold spell during those years in that region. Hence we can take the observations of the tree rings to be one method of "detecting" a cold spell during those years. But there is another possible explanation of those data. They might be due to the presence of a disease that affected the relevant trees during that period of time. Another kind of proxy data, therefore, could be used to rule out such a possible explanation and bolster the probability of the cold spell hypothesis being true. The two kinds of proxy data would then be RA diverse.

What about climate hypotheses "detected" by a simulation model? Suppose that a climate simulation can be used to calculate that equilibrium climate sensitivity (ECS) is greater than 2°C. One explanation of this is that ECS actually is greater than 2°C. Thus this would count as a detection of the hypothesis (that ECS is greater than 2°C) by a model. But another possible explanation might be that the calculated result is an artifact of the large grid size of the simulation. A natural move is to try to halve the grid size and check to see if the result is maintained. If it is, halve the grid size again. If the result remains stable, then the probability of that rival explanation goes way down. Thus, a reasonable ensemble of different simulation models with descending grid sizes could count as RA diverse. Context and judgment tell us when more and more of these stop adding to our credence in the hypothesis, just as it does in deciding how many different newspapers can continue to add to RA diversity regarding O.J.'s guilt.

But even once we are convinced that the grid size is not responsible for the purported detection of the hypothesis, there remains the possibility that the detection is an artifact of the way that cloud formation is parameterized in the simulation. A rival cloud parameterization can be tried. Certainly those two methods of detection would count as RA diverse. Again, context and judgment, but this time presumably of a more subtle and difficult character, would be required to decide at what point, if any, enough different cloud parameterization schemes are enough to rule out all such hypotheses.

All of the above suggests that the following notion of RA diversity might successfully explicate RA for a very wide range of hypotheses and detection methods:

> Explanatory RA-diversity: A set of detection procedures is RA-diverse with respect to some hypothesis H just in case the set can be put in some sequence $P_1, P_2 \ldots P_n$, such that for every k<n, detection procedure $P_{k+1}$ can be used to eliminate a competing hypothesis which itself could adequately explain $P_1 .. P_k$. (Schupbach 2016).

If we are right that explanatory RA diversity is the right notion of diversity for explicating all notions of robustness analysis, what does this tell us about the importance of robustness for climate science? A number of interesting results follow:

1) There is no single correct answer to the question: how much reliability do robustness considerations confer on climate hypotheses? It is a direct consequence of the definition of RA diversity we are considering that "adequate RA diversity" is not a property of a set of detection methods *tout court*. It is a property of a set of detection methods *with respect to a particular hypothesis and its rivals*.

2) The importance of the robustness of the results of climate *simulations* can now be more carefully assessed. What should we make of results that follow from most or all of our best global climate models (GCMs) and earth system models (ESMs)? The answer is that this is the wrong question. We should not ask whether the outputs that are robust under the ensemble of opportunity of models should be trusted. This

question is too simple to have a determinate answer. Rather, we should talk about specific climate hypotheses, and inquire about how RA diverse our ensemble of models is with respect to each one of these hypotheses individually. Furthermore, this is not simply a matter of deciding whether we have an antecedent commitment to the adequacy for purpose of one of the models vis-à-vis the hypothesis (*pace* Parker 2011). Instead, deciding whether a set of models is RA diverse for a particular hypothesis requires us to contemplate what competing hypotheses there are that might explain each model's detection of H.

3) Whether or not an ensemble of models is a good candidate for lending strong support for a hypothesis via RA depends almost entirely on the extent to which the set of models suffices for ruling out competing hypotheses. This means that just because the set of procedures we have that detect H are RA diverse does not imply that we should have confidence in H. RA diversity only implies CEP, i.e. it only implies that you are headed down the road to acceptance as you increase the size of the set of procedures. Once we know that a set is RA diverse the question of whether it is large enough to warrant acceptance of H, whether it is sufficiently RA diverse, is a further question. And the answer to that further question will always be a matter of judgment, context, considerations of inductive risk, etc.

4) Robustness analysis in climate science needn't be limited to examining the degree to which our GCMs and ESMs agree (though it can do that too!) – it can also be used to examine the degree to which hypotheses are supported by the whole range of methods available to climate scientists: modeling measurable data, modeling proxy data, and reasoning with equilibrium models as well as running GCMs and ESMs and other methods. Our best-supported climate hypotheses are the ones that are robust in this much wider sense – they are the ones that are at the confluence of multiple such methods.

## 12.3 Evidence of an ECS Hypothesis: Solely from Model Diversity

We don't have the space here to examine the full range of climate hypotheses supported by simulation modeling, and to do a robustness analysis for each one of them. Instead, we can look at one extremely important hypothesis, one regarding the true value of equilibrium climate sensitivity (ECS), and start by considering what would be involved in using RA to show that an ensemble of climate models provided robust evidence for a

---

**Box 12.1 Explanatory Power Formalized**

In giving the "logic" of RA, Schupbach gets a bit more formal then we have space to do here, and shows the probabilistic implications undergirding the explanatory details involved. For this he relies on a probabilistic conception of explanatory power according to which the explanatory power an explanans (h) has over its explanandum (e) is

$$\varepsilon(e,h) = \frac{Pr(h\,|\,e) - Pr(h\,|\,\neg e)}{Pr(h\,|\,e) + Pr(h\,|\,\neg e)}$$

As he shows in his paper, once one plugs this formal definition of explanatory power into the definition of explanatory RA, there is some payoff in terms of uncovering the normative upshot of RA – for example, when hypotheses compete in the sense of being mutually exclusive, the amount of confirmatory boost that the target hypothesis of an RA gets with each successful increment is determined by how epistemically good (with respect to prior plausibility and fit with the earlier evidence) the competing hypothesis was that just got eliminated in that last round. "Spoils go to the victor." Readers interested in the details regarding this should read not only Schupbach (2016) but also an earlier paper in which the above notion of explanatory power is defended (Schupbach and Sprenger 2011).

---

hypothesis regarding ECS. We can then move on to using RA to evaluate the full spectrum of evidence for such a hypothesis.

What would count, for example, as robust evidence that ECS was between 2.1°C and 4.7°C?:

Table 9.5 (IPCC, AR5, WG1, p. 818) of the IPCC AR5 presents a wealth of data regarding this question. In particular, it shows a range of values for ECS derived from 30 different models ranging from 2.1°C to 4.7°C, with a mean value of 3.2°C. This means that the hypothesis "ECS is between 2.1 and 4.7°C" looks superficially "robust" under all of our best models – our ensemble of opportunity. But is our set of models "RA diverse" enough to warrant confidence, all by itself, in that hypothesis? At the present time, we are not in a position to claim that it is. Let's reason through what it would take to employ RA to achieve such a result:

To begin, we need to recognize that ECS is the result of many different physical processes – in particular, in addition to the direct effect of carbon in warming the atmosphere, it is also the effect of many different climate feedbacks: water vapor feedback, lapse rate feedback, cloud feedbacks, albedo feedback, etc. But many climate scientists believe that cloud feedbacks are the main source of the difficulty in accurately predicting ECS with global climate models.[1] In other words, they believe that the unreliability of climate models vis-à-vis the resolution of cloud feedback mechanisms is the main reason that the models leave us with substantial uncertainty regarding the correct value of ECS. To put this in terms of the language of RA, this means that any "detection" of a range of values of ECS by an ensemble of GCMs and ESMs has at least two possible explanations:

1) ECS actually falls in that range
2) ECS does not fall in that range and the value predicted by the ensemble is an artifact of the systematic failure of all the models to accurately capture all of the feedbacks – with cloud feedbacks being an especially likely candidate.

Of course, we can check relatively easily whether our ensemble of models agrees with our best estimate – from observations – of the degree of cloud feedbacks that we find the recent observable past. Let us suppose that it does. Still, this could either be because some of our models are accurately modeling cloud feedbacks, or it could be because they just happen to be getting the recent past right *in virtue of compensating errors* – errors that they will not necessarily continue to compensate at the higher levels of forcing that we are interested in. Maybe, in other words, they are overestimating the degree of low-level cloud formation in the tropics, and underestimating the degree of high-level cloud formation in the higher latitudes, and these two mistakes are canceling out. But maybe when it comes to a regime for which do not yet have data, the degree of overestimation of the one will go up faster than the degree of underestimation of the latter. We have done nothing to rule out this alternative possibility.

What follows is this:

To count as RA diverse with respect to a hypothesis regarding ECS, the members of a set of climate models would have to be, at a minimum, sufficiently RA diverse with respect to a corresponding hypothesis about cloud feedback. And to count as sufficiently RA diverse with respect to a hypothesis about cloud feedback, the set of models would have to both correctly detect the degree of cloud feedback in the observable past, and

---

[1] A second major source of uncertainty regarding what a particular emissions scenario will do to the climate is the "climate-carbon cycle feedback." As the planet warms, it likely becomes less efficient at sequestering carbon. But this effect is often not included in the standard definition of ECS, because ECS is defined in terms of the response to a particular level of carbon. (And of course, that means that we might very well cause a doubling of the carbon level without having added that much carbon ourselves.)

they would have to systematically rule out, as possibilities, the various ways in which the models that are getting that right are doing so as a result of compensating errors. If we could achieve that, we would have a reasonably good argument – one based on RA – that there is a high probability that we are correctly detecting a hypothesis about cloud feedbacks in the future climate.

This explains the tremendous interest that climate scientists have in understanding the various features of climate models that give rise to the degree of variation in levels of cloud feedback that we find in our ensembles. In particular, as I will argue in this section, it explains their desire to uncover so-called emergent constraints[2] (EC) regarding cloud feedback and other climate feedback processes. Indeed, I would argue that EC reasoning is one of the very best kinds of RA reasoning. that we can find in climate modeling.[3] So let's take a minute to understand how EC reasoning in climate science works and then see how it relates to RA.[4]

Suppose you have a climate variable, v, that is of interest to you, (for example, the strength of some feedback mechanism – like melting snow causing less solar radiation to be reflected), and that variable has a plausible analog associated with seasonal variation or internal variability (for example, maybe that snow albedo mechanism that operates at the climactic level also operates from summer to winter). Call the strength of the seasonal mechanism w, and suppose you know the following facts:

1) If you run your whole ensemble of models over some very recent period for which there exists good data, you find that in that regime, your ensemble of models predicts values for w that lie in the interval (a,b). I.e. a<w<b.

2) Trustworthy recent observations of recent seasonal or multiyear variability in fact show that the value of w for the twentieth century lies somewhere in the interval (c,d), (and ideally, but not necessarily (c,d) lies entirely inside (a,b). I.e. a<c<d<b.

3) Systematic studies of all of your models reveal an emergent functional relationship (given by, say, f) between the intermodel variation in the value of w and intermodel variation in the value of v under various

---

[2] "Emergent" here doesn't mean anything spooky: it just means that a systematic relationship arises out of the dynamics, rather than being one that can be seen, by mere inspection, to be built into the governing equations.

[3] This last qualification is important: I am concerned in this section specifically with the question of what we can learn from *model* agreement. Other kinds of RA are applicable when we consider other lines of evidence.

[4] For more detail on EC see Klein and Hall (2015). I closely follow their discussion here.

forcings. (Students of emergent constraints call the quantity w observable in the current climate the "current climate predictor" or just "predictor" and they call the quantity, conditional on some hypothetical forcing, the "future climate predictand" or just "predictand." In this language, condition 3 is just that there is a stable function from predictor to predictand.)

What could I conclude from the above three facts? In a perfect world, I would conclude that the true value of v, under a candidate forcing, was very likely to lie in the interval f(c,d). In the vernacular: I would conclude that the predictand could be shown to be "constrained" to f(c,d). Why only "in a perfect world"? Because in order for this line of reasoning to be convincing, we would have to have some degree of confidence both that the ability of the ensemble of models to capture the predictor and that the systematic functional dependence embodied in f that we have uncovered is not merely the result of *systematically compensating errors*. There are various competing explanations for why my ensemble of models correctly encompasses the range (a,b), and for why the function f is observed to obtain, and those correspond to all the various ways in which these things could happen via compensating errors. According to EC reasoning, however, in order to conclude that the emergent constraint on the predictand lies in the interval f(c,d) we would like to meet a fourth condition:

4) There is a single plausible physical explanation of the relationship f between a predictor and the predictand – in terms of an emergent physical process – and we have ruled out potential competing explanations (such as compensating errors).

When conditions 1–4 are met, we can then argue that there is an emergent constraint on the predictand, and the true value of v is, with some degree of confidence, in the interval f(c,d).

Let's look at an example in the form of what some climate scientists view as one of the very strongest candidates for a well-detected emergent constraint: the constraint on the degree of snow-albedo feedback in the climate system. Snow-albedo feedback exists in the northern hemisphere on a seasonal and multiyear (due-to-internal-variability) scale. As the land warms, snow melts and absorbs more radiation. A strong linear relationship exists between 1) the degree of intermodel spread in snow-albedo feedback when our ensembles of opportunity of models are run for observable conditions and 2) when they are run as simulations of climate warming based on future forcings (condition 3, above). But we can give a fairly good estimate of the degree of snow-albedo feedback that has been in play over the observed period (because it can be estimated from satellite measurements, etc.). The plausible range of uncertainty is interior to the predictor interval (conditions 1 and 2, above). Good arguments, moreover, suggest that the same simple land-surface physics are the overwhelming source of model spread in 1) and 2) – i.e. in both the predictor and the predictand. So the available evidence supports the claim that a simple physical mechanism is responsible for the close correlation between this predictor and predictand (condition 4). Thus, we can conclude that the snow-albedo feedback is a genuine emergent constraint, and that we can use the observed linear relationship to extrapolate the range of uncertainty regarding snow-albedo feedback over the observed period to the period of the predictand. Luckily, this turns out to be quite a bit smaller than the degree of intermodel variability.

What are the arguments that support the claim that simple land-surface physics are the source of model spread? They are threefold. First, climate scientists find this to be a physically plausible mechanism. Second, by adjusting the land-surface physics in the models in the ensemble (by way of adjusting the parameter that controls the distribution of snow cover and vegetative canopy cover in the model) they are able to reproduce the model spread. And third, one can confirm that the various models get the snow albedo in a geographically similar way – when they produce more snow cover, they do it in the same places. This rules out the possibility that the average response is being achieved by compensating (geographically distributed) effects. When there is only one physical effect at work then geographic compensation of error is the only remaining kind there can be.

Unfortunately, the need to meet condition 4) places strong limitations on what kinds of variables are likely candidates for us to uncover emergent constraints. In particular, we should be very skeptical of a claim of having discovered an emergent constraint on a climate variable x – if x is *subject to multiple independent physical influences* (or even, to a lesser extent, if it is subject to a single physical influence that might manifest in different ways in different geographical locations). That's because if there are multiple independent influences, then it will be very difficult to eliminate the possibility (an alternative explanation) that the apparent detection of a systematic function between predictor and predictand by our models is the result of getting some of the independent influences wrong *in compensating ways*.

Suppose, for example, that my variable of interest is ECS itself. Is it plausible that we could directly uncover an emergent constraint on ECS? Probably not. As we discussed above, ECS is the result of the combined effects of multiple different feedbacks: water vapor, lapse rate, cloud, albedo, etc. And so any purportedly uncovered relationship between intermodel variation in a predictor and a predictand *could be the result of compensating errors*. In the language of explanatory-RA, there are still competing explanations of our purportedly robust findings.

Hence we should not regard any variable as a candidate for an emergent constraint unless we have good reason to believe that the functional relationship we are finding between predictor and predictand is *driven by a single physical process*. Unfortunately, this also means that even the rate of cloud feedback (just one component of ECS) is probably not a good candidate for the discovery of an emergent constraint. Cloud feedback itself depends on a variety of different independent feedbacks from many cloud types (e.g. high and low clouds, tropical and extra-tropical clouds, albedo and emissivity, etc.).[5] each of which is more than likely driven by a different physical process. In order to uncover the role of cloud feedback in driving ECS, cloud feedback itself needs to be broken up into a number of distinct emergent physical processes.

One possible candidate for an EC in the domain of cloud feedback is for one component of cloud interactions that contributes to it: the relationship between temperature and extra-tropical low-level-cloud optical depth. Climate scientists believe that they have a good handle on the single physical process that drives this very specific feedback, but the argument is less strong than in the snow-albedo case. More generally, to have confidence in the correct value for the entire net effect of cloud feedbacks, we would have to break it down into all of its components (of which extra-tropical low-level-cloud optical depth would be only one of very many) and find ECs for each and every one of them. Only in such a case would there be a line of reasoning that would rule out all other plausible explanations of what our models are detecting about cloud feedback rates. Work on this is still in relatively early stages – in part because not all the relevant functional dependences in the models having to do with temperature and extra-tropical low-level-cloud optical depth are well understood. (For example, it is not well understood "why models exhibit a negative relationship between cloud physical thickness and cloud top temperature in warm temperatures" Gordon and Klein 2014, p. 6058.)

In any case, for each component of cloud feedback, what we would like to show is that the functional dependence of that component is not only present in all the members of our ensemble of opportunity, but that

---

[5] (Klein and Hall 2015)

it is genuinely robust. How we do this in each case depends a bit on the details. The important point is that the argument that the relationship is the result of a single straightforward physical process is crucial to making that argument. This feature alone vastly diminishes the likelihood that the observational successes that the ensemble enjoys – with respect to the particular component of the feedback in question – are just the result of compensating errors. And that feature enables climate scientists to more systematically explore how robust the models are in this respect by turning on and off, and tuning up and down, one single physical process in the model, and by combining what they learn from the models with a reasonable degree of physical understanding of why they are seeing what they see.

The important point that I want to make here is that EC-reasoning is at bottom explanatory-RA-reasoning. It is at bottom an attempt to turn ensembles of opportunity into systematic explorations of rival explanations of their detection of strengths of various feedbacks.

If an ordinary ensemble of climate models, an "ensemble of opportunity" as they are sometimes called, gives a range of values for ECS, this does not by itself provide robust evidence that ECS lies in this range – no matter how well the models fit available data. That's because, by itself, the degree to which the set of models fits data does nothing to systematically eliminate competing possible explanations of each model's detection of this hypothesis. On the other hand, if an ensemble of models fits a data set with respect to a variable that can be shown to be the result of a single, unified physical process, then what it robustly says about that variable is much more trustworthy. And so, for that variable, if it detects a systematic functional relationship between something we can observe and something we are interested in predicting, then the likelihood of the detection being due to an alternative hypothesis is vastly diminished.

In other words, EC-reasoning does not show that our ensembles of models are sufficiently explanatory-RA diverse with respect to hypotheses about ECS, or cloud feedback rate, or anything of that kind. Instead, it aims to show that they are sufficiently explanatory-RA diverse with respect to a particular hypothesis about the functional dependence between a particular predictor and predictand. And if the variable in the predictor and predictand is driven by a single emergent physical effect (as it is for snow-albedo feedback, and might be for extra-tropical low-level-cloud optical depth feedback) then we can sometimes do this. We can then use the range of values for the predictor that we acquire from observation, and extrapolate that predictor to the predictand using the robustly supported functional relationship. If we could ever manage to do this for all the components of climate feedback, then we could add[6] these up and produce robust evidence for a range of values of ECS! In such an ideal scenario, we would be able to mount robust evidence because, for each component, we would be ruling out a sufficiently large class of rival explanations of the detected functional relationship by looking at the behavior of an ensemble of models.

This shows two interesting things about robustness and model ensembles. The first is that, with careful reasoning, even ensembles of opportunity can be mustered to provide robust evidence for climate hypotheses. We just have to choose our hypotheses carefully. *This is one reason it is so important to understand* (as Schupbach taught us) *that RA diversity is hypothesis-relative.* You can make an ensemble of opportunity RA diverse without altering the ensemble by adjusting your hypothesis. The second is that demonstrating model-RA-diversity with respect to the various components of climate feedback (and *a fortiori* of ECS) requires an understanding of the emergent underlying mechanisms that drive it. Interestingly enough, the knowledge required to underwrite this is very much in line with what Knutti calls "process understanding" (see Box 11.1). If this is right, then gaining process understanding is not necessarily a separate kind of epistemic activity from RA, and the two are complementary, rather than competing, accounts of how we gain confidence in model results. This is a happy result for us, since Knutti does not provide much explanation of why process understanding can lead to confidence in model skill – but our application of Schupbach's account of RA uncovers the logic of it.

## 12.4  Evidence of an ECS Hypothesis: Beyond Model Diversity

Given the relatively poor degree of process understanding regarding the role of the various cloud feedback mechanisms, and hence the relatively poor role that model diversity can (at least at this point in time) play in providing robust evidence for hypotheses about ECS, why does the IPCC make the following declarations?

Equilibrium climate sensitivity (ECS) is *likely* in the range 1.5°C to 4.5°C with *high confidence*. ECS is positive, *extremely unlikely* less than 1°C (*high confidence*) and *very unlikely* greater than 6°C (*medium confidence*).

To understand this, we need to look beyond the degree of RA diversity that exists in the evidence provided by our model ensembles and look

---

[6] Assuming they are linearly additive to a reasonable degree of approximation, which they probably are.

more carefully at the full range of methods that can detect these hypotheses. Here, I would argue, we can find a quite diverse set of methods. (There are three hypotheses being evaluated in the above: that ECS is >1, that 1.5<ECS<4.5, and that ECS<6.)[7]

Detection methods for detecting hypotheses regarding ECS can be grouped into three categories: those that rely on models, those that rely on instruments and instrument records, and those that come from paleoclimatological proxy data.

### Models

The most straightforward way to estimate ECS from climate models is the one summarized in table 9.5 of the AR5: this simply indicates that the range of values that our ensemble of opportunity predicts goes from 2.1°C to 4.6°C. But model-based estimates can be achieved in other ways – including the kinds of process-based constraints studies that we discussed in the last section.

### Instruments

Instruments and instrument records going back to the 1860s reveal a number of facts about how the climate has responded to a number of external forcings, including not only the increasing concentration of greenhouse gases produced by industrial activities over the period, but also volcanoes, the 11-year solar cycle, other aerosols, and ozone.

### Paleoclimate

Proxy data can be used to reconstruct temperature and atmospheric composition data for the last millennium, for the last glacial minimum, and for the last 420 million years, and all of these data samples have been used to try to estimate ECS.

Each of these detection methods is subject to some (if not all) of the following sources of uncertainty:

a) *Data uncertainty*: both instrument records and paleodata are subject to measurement uncertainty. Importantly, we have uncertainty regarding both the *strength of the forcing*, and the *quantity of temperature change*

---

[7] I follow Knutti and Hegerl's (2008) account of the science here.

that comes in response. Paleodata are of course more uncertain, and more so the farther back we go. But instrument data are also uncertain, especially with regard to the strength of past forcings (we don't have great data on carbon concentrations from the early twentieth century).

b) *Lack of equilibrium:* many of the above methods of detection work in the following sort of way – a volcano erupts and this produces a temporary forcing on the climate system by reducing incoming radiation. While the forcing lasts, the instrument records show how the temperature responds. The problem in most cases is that the planet does not have time to reach equilibrium before the forcing is taken away, so we are only "observing" a transient response. It is also important to note that the larger the true value of ECS, the greater this problem becomes. This is because simple models can be used to show that the larger the true value of ECS is, the more slowly equilibrium is expected to be reached, and, over shorter intervals, the smaller is the percentage of the full equilibrium response we expect to see. This means that methods that rely on observing non-equilibrium responses have a very hard time putting an *upper bound* on ECS.

c) *Lack of independence from internal variation:* over especially short periods of time, such as the ones afforded by the solar cycle and volcanoes, we also have the problem of teasing out what is the response (even if we concede that it might only be the transient response) to the forcing, and what is internal variation.

d) *Simulation model error:* obviously, simulation-model-based estimates can suffer from model error, including discretization error, poor parameterization, and the neglect of important sources of feedback.

e) *Different response from different kinds of forcings:* in principle, all forcings are created equal. It doesn't matter what method you use to add or subtract energy from the climate system, the equilibrium response should be the same. In the short run, however, not all forcings are the same because they don't all impact the planet homogeneously. Ozone and aerosols, in particular, tend to be localized around the poles, and hence can leave a somewhat distorted imprint on the climate record during non-equilibrium time intervals.

f) *Different base state:* as we delve further and further into the paleoclimatological record, we are dealing with a planet whose base state is further and further from our own, and which might in effect have a different true value of ECS from our own. So even exact knowledge of ECS based on how the planet behaved hundreds of millions of years ago may not be perfect for estimating what ECS will be for us.

If we are interested in doing robustness analysis on these various detection methods, then it is helpful to think of each of these sources of uncertainty (a-f) as alternative possible explanations of various hypotheses detections. Suppose, for example, that using instrument data associated with a particular volcanic eruption, we find that the data support the hypothesis that ECS is between 1.5°C and 4°C. We can count this as a method of detection for this hypothesis. Thus, to do RA, we would want to ask: in addition to the truth of the hypothesis, what other explanations are there for the fact that this method detects that hypothesis? We can then go through a-f and determine which ones provide such alternative explanations. Here, we can rule out a, d, and f. We can rule out a because the hypothesis is "wide" enough (spanning the range from 1.5°C to 4°C) to cover likely measurement error. We can rule out model error, because it is not a simulation-model-based detection. And we can rule out the possibility that the detection is due to the data coming from a different base state, because we are using data from the recent past. Thus, any second detection method that didn't suffer from these sources of error could help to rule out either b or e, and hence would count as RA diverse compared to our first method. Paleoclimate data is especially helpful here. That's because paleoclimate data are not especially susceptible to b, c, d and e. But they are especially susceptible to a and f. They are also somewhat susceptible to b if ECS turns out to be especially high. Thus, in conjunction with each other, volcano and solar-cycle data, combined with paleodata, are pretty good at ruling out alternative explanations of the detection of the hypothesis that ECS >1.5 °C, and hence already by themselves provide robust evidence of that hypothesis. If we add instrument records running from 1860 to the present alongside instrument records of heat-trapping gases for the period, we have a detection method that is similar to the volcano/solar-cycle variety, but which is less susceptible to b, not susceptible to c, but more susceptible to a. Since the response time in this data set is longer than the volcano and solar-cycle case, it can be used to provide stronger evidence against alternative explanations of failure to reach equilibrium (especially for slightly larger values of ECS), and very strong evidence against internal variability, but is more susceptible to data error.

What is of course interesting is that this set of detection methods is very good at ruling out alternative explanations for the hypothesis that ECS >1.5°C, but not very good at all at ruling out alternative explanations of the hypothesis that ECS< 6°C. One good way to see this is to ask: if ECS were greater than 6°C, what would explain all of our detections that it is lower? Unfortunately, it is not hard to come up with explanations: suppose, for example, there is a strong but as-yet-unaccounted-for positive feedback mechanism. Then we would not expect our models to correctly detect the high value of ECS, and we would not expect our instrument records to detect it either, because (being a high value of ECS) it would act too slowly for them to see it. We would probably only expect to see it in the millions-of-years-scale paleodata – but those data sets have enough uncertainty that they are poor at eliminating such a hypothesis (as well as the fact that they are susceptible to f).

## 12.5 Conclusion

Robustness analysis helps us to see why we have high confidence that ECS is greater than 1.5 °C, but lower confidence that it is less than 6°C[8], and virtually none at all regarding any hypothesis that is more fine-grained than 1.5 °C <ECS< 4 °C. Returning to the topic of Chapter 7, we can also see why the IPCC finds it useful to put probabilities on these hypotheses *combined with* a further estimate of confidence. Climate scientists seem to believe that each detection method puts a low probability on ECS < 1.5 °C *and* on ECS > 6 °C. Both are statistical outliers in each of the methods. But the RA we just performed on each hypothesis reveals that the ECS < 1.5 °C probability estimate is likely to be much more resilient – precisely because it is more robust. This is presumably why, when it comes to the claim that it is "very unlikely" that ECS > 6 °C, the IPCC claims only "medium confidence."

## Suggestions for Further Reading

Schupbach, Jonah N. and Jan Sprenger. 2011. "The Logic of Explanatory Power," *Philosophy of Science*. A very useful companion paper to the above, in which the relevant notion of explanatory power is articulated.

Knutti, Reto and Gabriele C. Hegerl. 2008. "The Equilibrium Sensitivity of the Earth's Temperature to Radiation Changes," *Nature Geoscience*. An elegant summary of the various independent lines of evidence for our best estimate of the planet's equilibrium climate sensitivity.

---

[8] The evidence has a small degree of robustness because if the hypothesis that ECS is greater than 6 °C were true, it would be a little bit of a coincidence that the correct value was falling at the very high end of both our simulation models and our long-span paleodata. But it would only be a minor coincidence. So this should not make us confident in our ability to rule such a hypothesis out.

Knutti, Reto. 2018. "Climate Model Confirmation: From Philosophy to Predicting Climate in the Real World." A paper on climate methodology and epistemology written for a philosophical audience by an IPCC lead author. Argues that process understanding is the key to making judgments about when models have various kinds of skill.

# 13

# *Social Epistemology*

## 13.1 Introduction

Epistemology is the study of knowledge and justified belief. In the last few decades, work in epistemology has increasingly focused on ways in which knowledge and justification are collective achievements, rather than merely the outcome of individual efforts. This work is often called "Social Epistemology."

Since science is one of our most important knowledge-producing activities, philosophy of science and epistemology are (or at least ought to be) very closely related areas of study. Indeed much of what goes on in philosophy of science can properly be called epistemology of science. And of course most science takes place in socially organized institutions. So we should expect developments in social epistemology to be relevant to the philosophy of science. And given much of what we have learned already about the collective efforts required to produce knowledge of our climate system, we should expect some of that to be especially relevant to the philosophy of climate science.

Epistemology generally, and the epistemology of science in particular, traditionally focused on the actions and achievements of individuals. Traditional epistemology focused on the beliefs of individual agents, and with a few notable exceptions[1], focused on the role that individuals' direct interactions with the objects of their beliefs have in justifying them. In the epistemology of science, the focus was always on individual scientific inquirers, and on how they, individually, gather evidence from the world in support or refutation of their scientific beliefs.

"Social Epistemology" can refer to any of the various ways in which the study of knowledge has broken with that tradition. When the kind of

---

[1] David Hume's discussion of testimony is one of the most famous examples. (Hume 2011 (1748), pp. 70–84)

knowledge under study is scientific knowledge, the work is often called the "Social Epistemology of Science." There are a variety of ways in which the general study of knowledge and belief can be "socialized." Goldman and Blanchard (Goldman and Blanchard 2016) call these the three branches of social epistemology. One is that epistemologists can shift the focus of their study from the ways in which individual agents gather evidence for their beliefs, to the ways in which these agents' beliefs depend for the justification (fundamentally) on the believers' interactions with other agents. A second is that epistemologists can shift their focus from the doxastic states of individuals to those of groups. And a third is that epistemologists can focus their attention on the consequences, for knowledge acquisition and justification, of adopting certain institutional arrangements for acquiring knowledge.

Philosophy of science, for its own part, first took a "social turn" in the 1970s following the publication of Kuhn's *Structure of Scientific Revolutions* (Kuhn 2012). The focus of that work was on investigating the extent to which the norms that govern scientific inquiry – for example, the norms of evidence evaluation; those for how to appraise the merits of competing explanations; or the ones for how to weight the importance of, e.g., simplicity against scope, in evaluating competing hypotheses – were themselves social norms: norms subject to variation as the result of social forces rather than being immutable cannons of rationality.

That kind of social epistemology of science won't concern us here. Instead, I will be more interested in highlighting the various ways in which all three kinds of social epistemology outlined by Goldman and Blanchard (a focus on social interaction as the locus of justification; on groups as the possessors of opinion and knowledge; and on the knowledge-producing consequences of social arrangements and institutions) should be expected to play an important role in investigating the epistemology of climate science.

We have already done a fair amount of this in previous chapters. The importance of shifting our focus from individuals to groups as the agents of scientific knowing was apparent not only in our discussion of probabilities and credences, but also in our discussion of the distribution of areas of expertise required for the construction of climate models (in Chapter 9). What I would like to do in this chapter is to examine, in turn, how all three branches of social epistemology outlined by Goldman and Blanchard are likely to be especially helpful for understanding climate science. The main goal will be to illustrate the claim that climate science is, in a thoroughgoing way, a socially organized kind of science, and that many features of its epistemology need to be surveyed at the social level in order to be properly understood. I will do this by discussing an example of each branch of social epistemology. We will start with the idea of focusing on social interaction as the locus of justification.

## 13.2  First Branch: Social Interaction

What do ordinary (non-philosophy-of-science) epistemologists talk about when they engage in this branch of social epistemology? Traditional epistemology has always discussed the idea of the "sources of knowledge" (Goldman and Blanchard 2016): perception, introspection, reasoning, memory, etc. Some of these sources are basic (presumably all of the above) and some are derived. Note-keeping might be a source of my knowledge, in that the only reason I know that you wore a blue shirt yesterday is because I wrote it in a book, but note-keeping is not a basic source of my knowledge. Everything I know via note-keeping can be reconstructed as something that I know via perception, memory, etc. I perceive what's in the notebook: I remember seeing your shirt yesterday and writing its color in the book; I reason that I would not want to deceive myself, etc. One of the central questions of this branch is the following: is testimony a basic or a derived source of knowledge? This question is not terribly important to understanding climate science. But another topic (one that Goldman calls a "spin off" of the first one) is: the problem of learning from the testimony of experts.

Regarding many technical topics, society is divided into experts and laypersons. On many such topics, laypersons are in no position to form their own opinions from scratch. Imagine, for example, trying to form your own opinion, without the help of any experts, about whether or not there is experimental evidence for supersymmetric particles. In such a case, and in many others, laypersons must appeal to experts when deciding what to believe. What should the layperson do, however, when there is disagreement among the experts? One possibility is of course to remain agnostic. It might be reasonable to remain agnostic about whether the possibility of supersymmetric particles has been ruled out by the evidence. But agnosticism is not always possible, especially when it comes to policy-relevant science. What should the layperson do when she has a question she needs an answer to, and the experts disagree? The literature recognizes three possible strategies: 1) the layperson can evaluate the arguments of the rival experts for herself; 2) she can measure which view is endorsed by the majority (or largest plurality) of the expert community; or 3) she can evaluate the track records of the respective experts.

In the domain of climate science, many scholars and policy wonks have devoted a fair amount of attention to demonstrating the existence of a scientific consensus about climate hypotheses, particularly the hypothesis of anthropogenesis.[2] Even Barack Obama, the President of the United States,[3] has entered the fray using this kind of metric. In May 2013, he tweeted "Ninety-seven percent of scientists agree: #climate change is real, man-made and dangerous." In 2015, speaking about the United Nations Paris Climate Change conference, he announced that "99.5 percent of scientists and 99 percent of world leaders" think that climate change "is really important." (Hollingsworth 2015).

All of this would suggest an implicit endorsement of strategy #2 above: when the layperson is confronted with conflicting expert testimony, she should accept the view endorsed by the majority of experts. (Or at least, it's an endorsement of this strategy when that majority is overwhelmingly large.) It also suggests, of course, that the reason *you* (assuming you are a climate layperson) should accept whatever climate hypotheses you ought to accept, is that the large (or overwhelming) majority of experts tell you that those hypotheses are likely true.

Not all philosophers of science accept this. Kristen Intemann (Intemann 2017), for example, has argued that focusing on these empirical findings (of, for example the percentage of scientists who agree with this or that claim; or the percentage of peer-reviewed articles that reach this or that conclusion) is predicated on several "problematic assumptions about the nature of consensus" (p. 189) and that reinforcing these assumptions could undermine, rather than increase, public trust in good climate science.

Intemann believes that all of these studies presuppose that agreement among experts (measured in various different ways) implies that that which they agree on is "trustworthy knowledge that should guide our beliefs and policy decisions" (p. 191). This, in turn, revolves around a cluster of different possible assumptions:

1) That agreement among experts is sufficient for reliable indication of trustworthy knowledge
2) That peer-reviewed articles are the marks of a reliable process of rigorous scrutiny
3) That unanimity is either necessary or sufficient for producing reliable opinion.

Intemann worries that all of these assumptions are problematic.

---

[2] (Oreskes 2004, Bray and von Storch 2010, Doran and Zimmerman 2009, Bray 2010, Rosenberg et al. 2010, Farnsworth and Lichter 2012, Verheggen et al. 2014, Stenhouse et al. 2014, Carlton et al. 2015, Cook et al. 2016)

[3] For a few more weeks at the time that I wrote the first draft of this.

### 1) Agreement Among Experts

Intemann argues that agreement by itself is insufficient because consensus can be reached for a variety of reasons:

> Agreement may occur, for example, as the result of accident, widespread misunderstandings, reliance on limited and homogenous methodologies, coercion, or the presence of widely shared values and interests that are likely to lead to systematic biases (Goldman 2001; Solomon 2001; Tucker 2003; Miller 2013). At one time, there was widespread agreement among scientists that smoking did not cause lung cancer. But, in part, this was presumably because they were a relatively monolithic group of scientists funded by the tobacco industry who had vested interests in particular outcomes. (Intemann 2017, p. 191)

Intemann correctly points out that in order for agreement among experts to be a reliable indicator of trustworthy knowledge, the agreement has to be the result of a well-ordered set of practices or procedures. And she also correctly points out that there is not even agreement among philosophers of science and epistemologists what characteristics this set of practices or procedures must have in order for it to count as well-ordered.

> Some have argued that a reliable consensus requires that agreement be achieved or caused by the employment of shared standards or epistemological norms and must undergo critical scrutiny (Longino 2002). Some maintain that the agreement must be based on a convergence of evidence that occurs by employing diverse methodologies(Oreskes 2007; Miller 2013; Stegenga 2016). Miller (2013) argues that even convergence of evidence is not sufficient to address concerns about potential biases. For a consensus to be reliable, participants must also be socially diverse with different values and interests so as to minimize the possibility that the agreement is being reached solely because of the influence of those values and interests (Miller 2013). (Intemann 2017, p. 192)

Advocates of sound climate policy based on the best scientific knowledge, she therefore argues, should be careful not to give the impression that the scientific consensus that exists is "mere agreement," since this is liable to make policy makers suspect that the agreement is the result of groupthink or conspiracy.

### 2) Agreement Via a Reliable Process of Peer Review

Intemann is equally critical of studies that report the high percentage of peer-reviewed articles that agree with respect to the core claim of anthropogenesis. She urges that there are several problems with taking consensus in peer-reviewed articles as the mark of reliability:

- Journal referees, journal editors, and granting agencies are biased in favor of widely accepted results, methods, and frameworks
- Researchers proposing novel or unorthodox results or methods tend to have to clear a higher bar for acceptance
- Confirmation bias can lead reviewers and editors to favor views they already accept.
- Absence or failure of blind review can cause prestige bias such that less-well-connected researchers' work (which is more likely to match the consensus) is more likely to be accepted.

### 3) Unanimity

Even unanimous, or near unanimous, agreement is not a desirable condition to use as a proxy for reliability, according to Intemann. This is both because the presence of disagreement goes hand in hand with careful critical evaluation of claims and arguments. Unanimity as a standard is also dangerous, according to her, because it fosters the possibility of manufacturing doubt. She also believes that insofar as some of the claims that we argued for regarding values and science in Chapter 9 are correct, then unanimity could be a sign that not all sets of values were being represented in the scientific community.

What should we make of these arguments? I take it we can all agree with Intemann regarding at least two key claims.

1) Consensus and agreement are only good guides to credibility and reliability insofar as the social community of scientists is appropriately structured or well-ordered. In the absence of a well-ordered scientific community, consensus could be the result of any number of forms of corruption, conspiracy, or herd mentality.
2) In any event, what really gives scientific claims the hallmarks of reliability are things like: the ability to make novel predictions; convergence of evidence for core claims; being able to provide the best explanation of available data; and the like. Facts about the community of scientists, such as their degree of agreement, are at best proxies for these more traditional, non-social, epistemological features.

But even if these two core claims are correct, it seems to me that the only thing that follows from this is that laypersons who are interested in evaluating the reliability of a scientific claim (such as, for example, the claim of anthropogenesis) have to ask themselves a key question:

*Which is easier for me to evaluate: the relevant social characteristics of the scientific community that puts forward the claim, or the quality of the evidence for the claim in the first place?*

We can concede the point that there are no agreed-upon necessary and sufficient conditions for a scientific community to be well-ordered, and still argue for the following: in the case of climate science, it is much easier for the layperson to ascertain for herself that the community of scientists is well-ordered (more on this below) than it is for her to ascertain for herself that the available evidence supports the claim. And if that is correct, then it is entirely appropriate for laypersons to put a great deal of stock in the kinds of empirical findings (regarding consensus, agreement, and the like) of which Intemann is critical.

Given all we have discussed in the book so far, it should be easy to make the case that the layperson is not well positioned to evaluate all the evidence in favor of any climate hypothesis. Much of that evidence, just for example, comes from complex simulation models, carefully reconstructed data and proxy data sets. No layperson is in a position to evaluate any of that evidence from scratch. So evaluating the evidence from scratch is particularly hard in climate science.

What about the opposite side of the coin? How hard is it for the layperson to evaluate the social-epistemological claim that the relevant scientific communities are well-structured enough for consensus to be a good mark of reliability in climate science? I would argue that while the former is especially hard in climate science, the latter is especially easy. Indeed, climate science enjoys a number of features that probably make it easier to establish the relevant social-epistemological claims.

1) Climate science is carried out by a variety of overlapping disciplines. So even if one of the disciplines participating were corrupt, this would likely be uncovered by a nearby participating discipline. So, for example, as we saw in Chapter 2, when the claims of climate modelers come into conflict with the scientists who appealed to weather balloons, a debate gets carried out in public until the two sides reach agreement. Scientific disciplines are not hermetically sealed, and epistemic corruption would be detectable by the layperson unless it extended into nearby disciplines as well.

2) Climate science is overseen by the IPCC, which publicly dissects, analyses, and summarizes all the relevant science and adjudicates what the properly peer-reviewed findings are.

3) Climate science is scrutinized by a highly skeptical and very well-funded opposition. For example: I can think of no other scientific discipline that has had a wide sampling of its email correspondence hacked and investigated. And as the RealClimate blog argued and as we noted back in Chapter 2, it is very interesting to note "*what is not contained* in the emails. There is no evidence of any worldwide conspiracy, no mention of George Soros nefariously funding climate research, no grand plan to 'get rid of the MWP,' no admission that global warming is a hoax, no evidence of the falsifying of data, and no 'marching orders' from our socialist/communist/vegetarian overlords."

In sum: I don't think we should minimize the difficulties faced by laypersons in evaluating scientific claims. It is difficult *both* for them to evaluate the evidence for themselves from scratch, *and* for them to evaluate whether the relevant scientific community is well-structured enough for consensus to be a mark of reliability. But in climate science, there are features that make the former task especially difficult, and features that make the latter task easier than in some other cases. I think, therefore, that it is particularly important in climate science to pay careful attention to what the various claims and hypotheses are that enjoy widespread agreement and consensus. It is also important, I think, to pay careful attention to the claims that the IPCC finds to be very likely and which it endorses with high confidence. Though nothing can completely mitigate the fact that science is always and everywhere fallible, and that climate science is no special exception, the lay public *can do no better* than to trust the IPCC's findings on these core issues. *There simply is no more reliable way* for a layperson to judge what it is reasonable to believe about the climate than to adopt roughly the degrees of belief that the IPCC recommends.

## 13.3 Second Branch: From Individuals to Groups

The second branch of social epistemology begins by noting how common it is to attribute mental states of various kinds (beliefs, desires, intentions, etc.) to groups such as armies, corporations, nation states, and the like. Social epistemologists in this branch are most centrally concerned with the question of when the beliefs of these sorts of entities attain the status of being justified, or of constituting knowledge. In the social epistemology of science, these kinds of questions can take on special characteristics. In Chapter 6, for example, we encountered the problem of saying what it is for a group to have a particular credence, or range of credences, in a scientific hypothesis. Here I would like to turn to a topic in social epistemology that has received little attention: group authorship.

Authoring is not the same as knowing, though it cannot be understood independent of its relationship to the epistemic activities responsible for producing knowledge. Although *knowing* is not an inherently social or communicative notion, authorship is by nature social and communicative, even where it is not collaborative. Unlike mere knowers, (nonfiction) authors are *accountable* for the content and accuracy of the information they produce. An author must be able to justify and vouch for the truth of her claims, defend her products when challenged, and retract her claims when she cannot defend them in light of criticisms or new information. An author might make false claims, at least so long as she remains accountable and takes proper responsibility for them. Inquiries that are backed up by faulty processes of knowledge production and acquisition yield authors that are less reliable sources of information. But such failings are not failures in the status of *authorship*. Failing to know is not the same as failing to author.

There are currently no good strategies for understanding or ensuring accountable authorship in the context of radical collaborative research. Huebner, Kukla and Winsberg (Winsberg, Huebner and Kukla 2014, Huebner, Kukla and Winsberg 2017) argued that neither the multiple-author nor the single-author model of accountability is viable in these cases. So, if the results that are reported by such collaborations are to be authored at all, they must be *group-authored*.

Huebner, Kukla and I (hereafter, HKW) contrast "radical collaborative research" with two kinds of collaborative research that do not pose a fundamental challenge to a traditional notion of authorship. First, we considered a type of collaborative research where a few authors work together to produce a single co-authored paper. Borrowing a term from Andy Clark (Clark 1998, p. 106) we called this "catch and toss" authorship. Second, we examined a kind of collective research that relies on a widely distributed form of information processing, but which retains a high degree of centralized control over the reported results. We used the first type of case to demonstrate a way in which research can be deeply social without compromising authorship; we used the second type of case to demonstrate a critical respect in which the distribution of information processing does not entail the distribution of authorship.

Let's begin with the kind of case most familiar to philosophers. When philosophers collaborate, they typically exchange ideas in a way that allows each author to retain epistemic authority over the resulting paper. In the most familiar form of philosophical collaboration, one author develops an idea (either in discussion, or as she writes her section of a paper). She then "tosses" it to another author who "catches" it, revises or extends it, and then tosses it back. This type of collaborative writing often consists in a long series of such "catchings" and "tossings," and the process is typically repeated until every member of the collaboration is happy to put her name on the resulting paper. These types of collaborations allow every author to be accountable for her own contributions. But, more importantly, since the writing and thinking are genuinely *interactive,* each author is likely to form a relatively clear sense of the structure of the overall argument. While some minor disagreements might remain regarding what ought to be said, each author remains in position to vouch for the finished product. Each author is typically in a position to defend the claims that are produced through the collaboration when they are challenged, and each author is in a position to retract the claims that cannot be defended in light of criticism.

Of course, things are not generally this simple, and there is a different kind of "catch and toss" process that arises when a project requires a unique technical skill or information that is only available to a specialist. In some cases complex statistical analyses are required that a lead author does not know how to carry out; in other cases, the relevant research may require a working knowledge of some range of scientific data that is only available to a person who has undergone special training. But, in these cases, a single author can do the work that is relevant to a particular section of a paper in a way that allows her to retain responsibility for her portion of the paper. For example, one author may "catch" a data set collected by someone else, carry out a complicated statistical analyses, and "toss" these results back to the lead author who inserts them into the paper. Clear lines of responsibility are maintained in this type of co-authorship, and where the corresponding author is unable to vouch for a claim that is made in the paper she will know which collaborator can in fact vouch for it. This means that it is possible for collaborations to include specialists, while still making sure that there is someone who is epistemically accountable.

In both of these cases, "catch and toss" collaborations help to organize information in a way that functionally distributes the labor required for producing some purported piece of knowledge. The ideas produced through "catch and toss" collaborations are often only possible because of the patterns of reciprocal feedback that arise between collaborators. However, each part of the co-authored paper is the product of an individual author who produces representations that can be "tossed" back and forth. This yields a sort of joint authorship in which claims to knowledge are produced through a highly interactive process that leaves standard structures of accountability and responsibility in place.

There are types of research, and accompanying forms of collaboration, that are too complex for this "catch and toss" model to be practically possible. For instance, after the French Revolution, the Académie des Sciences famously instituted a metric system of weights and measures. They also stipulated that right angles would now be divided into 100 grades rather than 90 degrees. Nautical navigation relied heavily on trigonometric tables, and using this new standard of measurement on the fly proved practically impossible. So, the Académie instituted the Bureau du Cadastre to construct new trigonometric tables using grades rather than degrees. The director of the Bureau, Gaspard de Prony, was enamored of Adam Smith's idea of "distributing labor," and he hired a team of human computers to carry out the arduous task of constructing the trigonometric tables. The vast majority of these computers (around 90 of them) "were former servants and hairdressers, who had lost their jobs when the Revolution rendered the elegant styles of Louis XVI unfashionable or even treasonous" (Grier 2005, p. 36). As such, they had no mathematical training beyond the basic abilities required to add and subtract. So, a team of eight "planners" supplied these computers with worksheets that allowed them to carry out simple calculations; the planners then took differences between the calculated values to spot-check for errors. Finally, a small number of trained scientists figured out the appropriate formulas to be passed down to the planners. All of the relevant calculations were eventually passed along to de Prony, who put together a 19-volume manuscript that included all of the tables.[4]

Structurally speaking, this kind of research shares much with the "catch and toss" collaborations discussed above. However, in this case, the human computers became nothing more than sources of information for de Prony's mathematical investigations. The organization and distribution of intellectual labor instituted clear lines of responsibility, such that de Prony (and perhaps his team of trained scientists) retained accountability for the resulting tables. The human computers became mere tools, to be used in the service of a project carried out by a few highly trained and highly intelligent scientists. Because of the organization of this research, de Prony (and perhaps the trained scientists) remained epistemically accountable. These authors could vouch for the results of this massive collaboration; they were

---

[4] Unfortunately, the manuscript was never published because the publisher went bankrupt, and the Napoleonic government had no interest in publishing the volume.

the ones who were epistemically accountable for producing accurate tables, defending them if challenged, and revising them if necessary. The relevant structures of epistemic responsibility and accountability were maintained in the Bureau du Cadastre to allow for the distribution of collective research while leaving traditional notions of authorship intact. In effect, the text was still single-authored, even in the face of distributed information processing, because one person retained centralized control over the research process, including its methodological standards and implementation. While many people participated in the production of knowledge, only one person had the status of the author of the document communicating that knowledge.[5]

In the type of case we just examined, labor is distributed, but there is no need for lower-level agents to exercise special epistemic skills or judgment, because the author who retains centralized control can provide lower-level agents with detailed and specific instructions about how to handle every problem that they might encounter.

We now turn to a set of cases involving distributed epistemic labor, in which the primary reason for involving a large number of actors is that no single actor *can* possess the relevant knowledge and skills that are necessary to *produce and sanction* the desired knowledge claims. In such cases, multiple actors must exercise special epistemic skills and judgment. In the types of collaborations that interest us here, there is no one who understands the role that is played by every researcher, and there is no one who knows what everyone else has contributed to the project; there is no one who has even testimonial knowledge that the other researchers are trustworthy; there is no one who has command over how the various pieces of the study fit into a coherent whole.

HKW argued that the reports that result from these types of radical collaborations cannot be understood as multiply authored, as in the catch and toss model, or single-authored while involving distributed labor, as in the centralized control model. Instead, if they have any kind of author at all, it must be a *group author*.

No case exemplifies radically distributed epistemic agency better than climate science – especially when it involves massive, modular, and highly complex *coupled atmosphere-ocean global climate model* (AOGCM) simulations, such as the National Oceanic and Atmospheric Administration's GFDL CM2.X Model.

---

[5] Strikingly, a similar model of highly distributed, centrally controlled research was instituted at the Oak Ridge National Laboratory as a way of isolating isotopes of uranium-235. For a detailed discussion of this case see Winsberg, Huebner and Kukla (2014).

The sheer size and complexity of such models (see Box 4.3 for a reminder of some of the features of this complexity) should make it clear why their construction and sanctioning must involve a vast army of specialists. Their production and use requires expertise in climatology, meteorology, atmospheric dynamics, atmospheric physics, atmospheric chemistry, solar physics, historical climatology, geophysics, geochemistry, geology, soil science, oceanography, glaciology, paleoclimatology, ecology, biogeography, biochemistry, computer science, mathematical and numerical modeling, statistics, time series analysis, and more. Furthermore, no GCM is built from the ground up in one short, surveyable unit of time. They all rely on assemblages of methods, modules, parameterization schemes, initial data packages, bits of code, coupling schemes, and so forth, that have been built, tested, evaluated, and credentialed over years or even decades of work by climate scientists, mathematicians, and computer scientists of all stripes.[6] This yields a sort of "fuzzy modularity" in these models (Lenhard & Winsberg 2010). (See section 4.4.)

The problems for authorship that are evoked by this type of fuzzy modularity are exacerbated by the fact that epistemically salient decisions in climate modeling must be sensitive to a continuous, dynamic flow of information through the model. The collaborators must carry out continuous deliberative adjustments in light of new circumstances and new information, and this must frequently occur without the time or means to consult with other members of the research team. This is why the operation of each module must rely on a mixture of principled science and decisions about parameterization. Climate modeling involves literally thousands of methodological choices. Many crucial processes are poorly understood; many compromises must be made in the name of computational exigency; and so forth.

Meanwhile, climate experts, in light of the individually limited role that they play in the socially extended activity of building climate knowledge, can only arrive at opinions about the future of the climate in ways that are fundamentally mediated by the complex models that they build. And they are incapable of sorting out the ways in which past methodological choices made by other scientists – scientists whose expertise they don't entirely share – are influencing, through their entrenchment in the very

---

[6] There has been a move, in recent years, to eliminate "legacy code" from climate models. Even though this may have been achieved in some models (this claim is sometimes made about CM2), it is worth noting that there is a large difference between coding a model from scratch and building it from scratch, that is, devising and sanctioning from scratch all of the elements of a model.

models that mediate their inferences, the conclusions that they deliver to policy makers.

No single person is in a position to offer a rational reconstruction of a climate model. Too many of the methodological choices are buried in the historical past under the complexity, distribution, and historically path-dependent character of climate models (See Chapter 9). They might very well have been opaque to the actors who put them there, and they are certainly opaque to those who stand at the end of the long, distributed, and path-dependent process of model construction. There is no one who could, even in principle, be held accountable for the claims to knowledge that are produced using CM2.X. To put the point another way, there is no person who has the requisite expertise to play the role of de Prony in these modeling activities, and accordingly there is no single person who can legitimately be treated as the author of these claims.

HKW make a variety of claims about authorship in radically collaborative research that are beyond the scope of this chapter. The central claim that concerns us here is that if authorship exists in climate science, unlike in the catch and toss cases or in the de Prony-type cases, it is necessarily group authorship. And there are interesting puzzles about group authorship that need to be sorted out if we want to understand authorship and accountability in climate science. Like what?

## 13.4 Third Branch: Institutional Arrangements

Virtually everyone agrees that dissent is a crucial feature of the enterprise of science. The people who propose novel ideas, hypotheses and theories are far less likely to identify problems with their own creations than are their critics. So the ability of the scientific community to self-correct – to improve on partially correct ideas and theories and to replace more deeply flawed ideas with new ones – depends for its success on critical exchange between proponents of these ideas and dissenters.

But at least since the publication of *Merchants of Doubt* (Oreskes and Conway 2010), some have come to realize that not all instances of dissent are good – neither epistemically good for the growth of knowledge, nor ethically good for the improvement of society as a whole. In their book, Naomi Oreskes and Erik Conway argue that a small group of scientists backed by powerful interests "manufactured controversy" about such topics as the health risks of smoking, the dangers to the ozone layer, the effects of acid rain, and of course the existence of anthropogenic climate change. Their well-documented claim, in a nutshell, is that this small group of scientists used millions of dollars of funding to create a kind of illegitimate dissent, doubt, and controversy.

If Oreskes and Conway are right, and by all accounts they are, then there must be a distinction between legitimate dissent (which is fundamental to science) and "manufactured" dissent, which is both epistemically detrimental to science and ethically detrimental to society. But what, exactly, makes dissent illegitimate or detrimental, and what institutional structures that promote dissent are similarly illegitimate? (For an apparent example of the latter, the American Enterprise Institute (AEI) offered a $10,000 prize to scientists who found evidence that contradicted the AR4 of the IPCC.) This question has been asked by Justin Biddle and Anna Leuschner (Biddle and Leuschner 2015), and it falls squarely in the third branch of social epistemology, since it asks: when does a kind of social structure – a monetary prize for achieving a scientific result – a piece of social structure that is normally taken to promote the epistemic goals of science – turn into a wrong: something that hinders rather than promotes those goals? It turns out to be somewhat more difficult to answer this question than one might anticipate. This difficulty becomes apparent when they survey a variety of possible answers.

i)   First answer: the AEI prize is problematic because it rewards research that achieves a particular result, rather than rewarding the finding of truth, come-what-may.

Biddle and Leuschner reject this answer because, they argue, it is perfectly normal for scientists to aim to attain specific results. This is surely right. If we learned anything at all from Thomas Kuhn (Kuhn 2012 (1962)) it is the degree to which scientific activity is devoted to bringing nature into conformity with accepted knowledge. Think of the tremendous efforts expended to find the Higgs particle, or to experimentally detect cold dark matter, etc. Surely it would not be intrinsically epistemically detrimental to award a prize for anyone who provides laboratory evidence of cold dark matter.

ii)  Second answer: the AEI prize is problematic because it is motivated by a political and economic agenda.

Biddle and Leuschnerreject this answer by appealing to one of the first works on the social epistemology of scientific institutional arrangements, Philip Kitcher's *The Advancement of Science* (Kitcher 1993). In the book, Kitcher argues that impure motivations often promote good research by helping to distribute the epistemic labor of individual researchers over a whole space of possible research strategies. If James Watson is to be believed, for example, he pursued the DNA hypothesis over the protein hypothesis not because he thought it was true, but because it was unpopular, and he therefore assumed that *if it were true*, it would be more likely to make him famous if he worked on it. So the fact that Watson was motivated by his desire for fame was epistemically beneficial to the community as a whole. If Kitcher is right about this, then it can't be a problem by itself that the AEI prize is motivated by a political and economic agenda. Maybe, after all, the existence of that agenda simply serves to diversify scientific opinion in an epistemically beneficial way.

iii) Third answer: the AEI prize is problematic because it is motivated **only** by a political and economic agenda.

In the same work (1993) Kitcher argued that scientists with "grubby" motives (desire for celebrity, credit, monetary reward, etc.) only promote scientific progress when those motives are *combined with* a desire to seek pure epistemic goals like truth or reliability. If "grubby" motivations are unaccompanied by epistemic motives then let us call them "depraved" motivations. It is plausible to think that depraved motivations are epistemically detrimental since they will encourage "wishful thinking" on the part of scientists: deliberately skewing their methods, interpretations, etc., in favor of the desired result. If Kitcher is right about this, then maybe what is wrong with the AEI prize is that it is designed to induce scientists to entirely abandon epistemic goals and it encourages the development of depraved motivations.

Biddle and Leuschner concede that if the AEI prize fosters climate skeptics with depraved motivations then that is sufficient grounds for judging it to be epistemically detrimental. But they have two worries. The first is that while depraved motivations might be a sufficient condition for showing that a prize is epistemically detrimental, it might not be a necessary condition. It might be that climate skeptics with a sincere belief in the falsity of anthropogenesis are, despite their sincerity, pursuing epistemically detrimental work. Their second worry is that this criterion requires us to engage in a kind of mind reading in order to determine who is engaged in detrimental dissent: we have to uncover scientists motivational structure.

iv) Biddle and Leuschner's answer.

Biddle and Leuschner give jointly sufficient conditions for determining whether dissent is epistemically detrimental. Before examining these, however, it is crucial to note that Biddle and Leuschner reject the value-free ideal of science that we discussed in Chapter 9. In particular, they reject the claim that methodological choices can be made in a value-free way. One might naturally think that illegitimate dissent is simply dissent motivated by a particular set of non-epistemic values. But because Biddle and Leuschner reject the value-free ideal, this answer is not available to them. One who rejects the value-free ideal thinks much legitimate research is guided by one set of values or another. To get around this problem, they appeal to the work of Torsten Wilholt (Wilholt 2009), who faced a similar problem in his attempt to articulate the conditions under which research can be said to suffer from "preference bias". "Preference bias" prima facie, is when researchers are predisposed in favor of one possible answer to a question under investigation, and skew their research toward achieving that answer. But as a skeptic of the value-free ideal, Wilholt maintains that methodological choices are often impossible to make in a value-free way. So for him, as well, a plausible answer is not available: that preference bias simply exists whenever researchers make a methodological choice that can only be motivated by a desire to get one answer rather than another. To the critic of the value-free ideal, this happens far too often to count as bias. Instead, he argues that preference bias "is the infringement of an explicit or implicit conventional standard of the respective research community in order to increase the likelihood of arriving at a preferred result." (Wilholt 2009, p. 99)

What does Wilholt mean by a conventional standard? This is best explained with an example.

> The use of a strain of rat that is particularly insensitive to estrogen violates a conventional standard in toxicology that was made explicit by a group of experts convened by the U.S. National Toxicology Program: "Because of clear species and strain differences in sensitivity, animal model selection should be based on responsiveness to endocrine active agents of concern (i.e., responsive to positive controls), not on convenience and familiarity" (quoted in Wilholt 2009, p. 97)

In mirroring Wilholt, Biddle and Leuschner are led to adopt the following set of jointly sufficient conditions for counting scientific dissent as epistemically detrimental.

Dissent from a hypothesis H is epistemically detrimental if each of the following obtains:

1) The non-epistemic consequences of wrongly rejecting H are likely to be severe.
2) The dissenting research that constitutes the objection violates established conventional standards.

3) The dissenting research involves intolerance for the risk of accepting H.
4) The risks of accepting H and being wrong and the risk of rejecting H and being wrong fall largely upon different parties.

I agree with much of what Wilholt says about preference bias and the role of conventional standards. And thus I would also suggest that there is much to recommend about Biddle and Leuschner's analysis. The problem in both cases, however, is that they both strike me as being overly optimistic about the scope of conventional standards. Wilholt identifies one case of preference bias in which a conventional methodological standard was violated. And Biddle and Leuschner do the same for a case of detrimental dissent. But it seems far from clear to me that there exists a detailed enough set of conventional standards in the world to do the relevant kind of policing that these authors hope they will do. Take the example that Biddle and Leuschner give of a violation of a conventional standard by a climate dissenter. They claim that research by McIntyre and McKitrick violated the standard that one should "ensur[e] that the data that one tosses out is noise and not signal." (p. 266) It is of course true that this is a genuine standard. But it is far less clear that there are appropriately fine-grained conventional standards for, in turn, ensuring that data are not noise. Recall all the difficulties we surveyed in Chapter 2 that are often involved in constructing good models of data. There are no simple rules to follow. And if there are no conventional rules for constructing data sets, then the conventional rule that one ensures that the data one rejects be noise becomes rather empty.

On the other hand, I also worry that Biddle and Leuschner are unduly pessimistic (in their discussion of the third answer above) about our ability to uncover the motivational structures of scientists. I suspect it is not as hard as they make it out to be to uncover cases in which dissenting scientists have "depraved" motivations. There can be various sources of evidence of what people's motivations are, even if they are not completely transparent to us. We can find traces of them in various kinds of historical records, communications, funding sources, and the like. Thus, I am somewhat unconvinced that focusing on depraved motivations is a bad way to identify detrimental dissent, and I am also unconvinced that it is a bad idea to investigate whether a prize like the one offered by the AEI is fostering depraved motivations in deciding whether, as a piece of institutional scaffolding for fostering knowledge production, it is beneficial or detrimental. Perhaps the worry is only this: if "depraved motivations" are the hallmark of detrimental dissent, then it will need to be historians or investigative journalists, rather than philosophers, who will be in the right position to uncover detrimental dissent.[7]

## Suggestions for Further Reading

Goldman, Alvin and Thomas Blanchard. 2016. "Social Epistemology," in *The Stanford Encyclopedia of Philosophy*. An introduction to social epistemology. Outlines the "three branches" that we discuss in this chapter. Contains an extensive bibliography.

Kitcher, Phillip. 1993. *The Advancement of Science: Science Without Legend, Objectivity Without Illusions*. One of the first works of the social epistemology of science. A classic in the field.

Intemann, Kristen. 2017. "Who Needs Consensus Anyway? Addressing Manufactured Doubt and Increasing Public Trust in Climate Science," *Public Affairs Quarterly*. Argues against the usefulness of empirical findings regarding the degree of consensus over various climate hypotheses.

Huebner, Bryce, Rebecca Kukla and Eric Winsberg. 2017. "Making an Author in Radically Collaborative Research," in *Scientific Collaboration and Collective Knowledge*. Argues that a useful notion of "group authorship" should be one that focuses on the existence of a group level entity that is *accountable* to challenge and criticism. The paper focuses on three case studies, one of which is climate science.

Biddle, Justin B. and Anna Leuschner. 2015. "Climate Skepticism and the Manufacture of Doubt: Can Dissent in Science Be Epistemically Detrimental?," *European Journal for Philosophy of Science*. Explores the question of when, if ever, scientific dissent is epistemically harmful. The authors argue that this situation obtains when dissenters violate conventional scientific standards.

Wilholt, Torsten. 2009. "Bias and Values in Scientific Research," *Studies in History and Philosophy of Science Part A*. Introduces the idea of a conventional scientific standard as the solution to a problem that arises once we recognize the role that values play in science. Provides the philosophical background for Biddle and Leuschner, above.

---

[7] As, for example, was done by Oreskes and Conway (2010)

# 14

# *Epilogue*

I opened the book with, among things, a plea for a proper appreciation of the richness and complexity of climate science. The best-supported claims of climate science, the ones to which the IPCC assigns high probability with high confidence – that there is a detectable externally forced warming trend, that the trend is attributable to human activities, that ECS is between 1.5°C and 4.5°C, etc. – are supported by decades of painstaking work in collecting and studying data, pursuing multiple independent lines of evidence, and building and studying complex models run on clusters of powerful computers.

One reason for making this plea was so that philosophers would come to appreciate the degree to which the conceptual, methodological, and epistemological issues that perennially preoccupy philosophers of science come to life in various interesting and novel ways in climate science. If you've made it this far in the book, I hope it is because you've come to agree with that claim.

The relationship between hypotheses and data in climate science is more complex than some accounts make it out to be, and misunderstandings about this sometimes lie at the root of misunderstandings, more generally, about how well the climate system is understood. Models and simulations in climate science should be understood primarily as pragmatic tools, rather than as generally confirmed representations. They should be evaluated, first and foremost, as either having or lacking relevant kinds of "skill," rather than as adequately representing the world, in some generalized kind of way. The principal kinds of skill we want them to have is to be good at making projections, rather than predictions. Failure to appreciate that difference is another source of misunderstanding about the epistemic power of climate models and simulations; especially in the face of the chaotic character of the atmosphere and the other systems to which it is coupled.

The right way to understand claims, particularly those made by the IPCC, to the effect that some hypothesis or other is "very likely," or "virtually certain," or "more likely than not," is as credences. Even though these are often the official pronouncements of groups of individuals to which we want to attach the greatest possible degree of objectivity, we nevertheless understand them as statements of personal probability. Even though these are often the official pronouncements of groups of individuals to which we want to attach the greatest possible degree of objectivity, we nevertheless understand them as statements of personal probability; or credences. While the basic mechanisms governing the climate system and anthropogenic climate change are very well understood, significant uncertainties remain concerning the question as to how *precisely* the climate system will change, especially on regional scales. In fact, our epistemic situation about the future of the climate system and its effects on social and economic systems displays two levels of uncertainty. We are not faced with outcomes with known probabilities, that is, but rather with situations in which we are not epistemically warranted in positing a probability distribution over sets of possible outcomes with high confidence. Values enter into science precisely in situations of complex modeling adaptable to a variety of purposes, and in situations of risk and uncertainty.

We have also surveyed several issues related to the epistemology of climate science – to establishing the skill of particular climate models, the credibility of particular climate hypotheses by looking at ensembles of climate models, and finally of establishing the credibility of climate hypotheses by assembling independent lines of evidence including but not limited to climate simulation models. Despite the ubiquity of the "verification and validation" framework for thinking about the epistemology of simulation, it is poorly adapted to the application of simulations that rely heavily on parameterizations and tuning. It is also poorly adapted to situations, such as the one we face in climate science, where the software modularity of our models overstates the degree of scientific modularity of the underlying system. Despite some of the difficulties philosophers have faced in properly analyzing the underlying concept, climate hypotheses that are supported by both *robust* lines of evidence and *robust* ensembles of models are the ones in which we should place the most trust. Being able to identify the characteristics that ensembles of models ought to display in order to count as genuinely robust is not at all unrelated to having process understanding of the system being modeled.

Finally, we looked at epistemological issues that relate specifically to the social characteristics of climate science: what should non-experts (including philosophers of science) infer from the fact that there is a consensus of experts regarding some climate hypotheses; what is the nature of authorship in a radically epistemically distributed science like climate science; and given the usual assumptions about the value of dissent in science, what can we say about when scientific dissent begins to become detrimental (as sometimes seems to be the case regarding the hypothesis of anthropogenesis)?

All of these discussions and lines of argument have profound consequences for how we ought to think about the epistemic status of the main claims of climate science. Consider what we learned, in Chapter 2, about the nature of the relationship between models and data. The complexity of this relationship explains, in part, why it has been so easy for climate change deniers to manufacture controversies like the troposphere controversy (see section 2.4) and the climategate controversy (Box 2.3). It also underlines why it is dangerous for laypeople to think they are well positioned to evaluate, using ordinary reasoning and common sense, the strength of the arguments behind these controversies. As we saw in section 2.4, settling the question of what the troposphere data showed about the quality of climate models was not easy, and required the efforts of scientists with diverse backgrounds and areas of expertise.

Or consider what we learned about models and simulations, and their use in studying complex and chaotic systems, in Chapters 3, 4, 5 and 10. While less dogmatic than outright climate change deniers, there are some milder critics of climate science who believe that models and simulations shouldn't be understood to contribute much epistemic weight to our confidence in many climate hypotheses. Whether because of general doubts about computational methods, or because of worries about the chaotic nature of the atmosphere – worries about butterflies and moths – they believe that only the results of "basic science" should be relied on; especially when adopting policy recommendations. Much of what we learned in the above chapters can be helpful, I think, in putting to rest many of these doubts. And what we learned in Chapters 3, 11 and 12 also helps us to understand the ways in which models and simulations function in harmony with other sources of evidence in securing the claims for which the IPCC has the greatest confidence and to which it assigns the highest probabilities.

And finally, in Chapter 13 we learned some useful things about what epistemic attitudes that we, as climate laypersons, should adopt about findings for which there is broad consensus in a scientific community: while we should always be mindful of the fact that the findings of science are fallible, there are circumstances under which it is plainly irrational for laypersons to try to second-guess the results of our best science. And we learned that, while criticism and dissent are among the most important hallmarks of science, there are circumstances under which scientific dissent can be epistemically harmful.

I want to end the book by devoting a small amount of space to a topic we have not discussed very much. A big part of the project of this book has been to get a philosophical grip on how climate science works – on the methods by which it establishes that either full belief, or high degrees of belief, in certain hypotheses is justified. Because of that, I have focused primarily on the sanctioning of claims that enjoy that degree of justification: that there is an observable pattern of warming over the last century and a half; that this pattern of warming is attributable to human activities that increase the concentrations of heat-trapping gases in the atmosphere; that the equilibrium climate sensitivity lies somewhere in the range of 1.5°C to 6°C; etc.

But I don't want to leave the reader with the impression that establishing high degrees of belief in climate hypotheses is the only policy-relevant thing that climate science is good for. This is a second important reason for emphasizing the richness and complexity of climate science and discouraging the view that climate science is like 1+1=2. In particular, I also want to urge that we take seriously the idea that climate science can be informative in another kind of way: by raising the *possibility* of various outcomes. I especially want to urge this with respect to claims regarding low probability outcomes, and even claims regarding outcomes for which it is impossible to attach a range of probabilities with any degree of confidence. These can be outcomes, nevertheless, which we should take seriously because of their potentially catastrophic consequences. Many such outcomes fall under the description of what is sometimes called "abrupt climate change."

The climate system is a complex, chaotic, multivariate system that is almost certainly capable of exhibiting, under the right circumstances or forcings, behavior that is qualitatively and radically different from anything we have seen before. We can see examples of what would give rise to this in the geological record: cases where a moderately sized change in one component of the climate system pushed the system as a whole into a radically different regime. The very same sort of thing *could* happen as a result of our pushing greenhouse forcings past some presently unknown threshold. Not only might we trigger abrupt and disruptive changes in the climate, but changes might be relatively irreversible once they are set in motion. Nothing, moreover, guarantees that changes of this kind will not happen over a very short period of time – decades or even years.

Possible abrupt changes of this kind include disruption of the Atlantic meridional overturning circulation (AMOC), rapid release of methane from the Arctic sea floor, ecosystem collapses, collapse in the Arctic sea ice, collapse of ice sheets in Antarctica or Greenland, and release of methane from the melting of the Arctic permafrost. All of these sorts of occurrences are considered very unlikely over the next century. But at the same time, most of those probability assessments are made with very low confidence. Part of the reason we are stuck having low confidence in these assessments is precisely that we have no data that are relevant to these regimes, and so it is nearly impossible to accumulate independent lines of evidence in support of them or otherwise. The right epistemic thing to say about them, perhaps, is that we should simply remain agnostic about them for the time being. But all of them would have catastrophic consequences. Our best philosophy of science, formal epistemology, and decision theory are, however, poorly adapted to thinking through how to act rationally in the face of that kind of epistemic situation. Some of the most important future work in the philosophy of climate science and climate policy, I would wager, will be directed at this class of problems.

# Appendix: Structural Stability and the "Hawkmoth Effect"[*]

## A.1 What Is the Hawkmoth Effect Supposed to Be?

In a series of recent papers,[1] Roman Frigg, Leonard Smith, Erica Thompson, and others,[2] warned of the skeptical implications of what they have termed "the hawkmoth effect." Nowhere do they precisely define the hawkmoth, so we will have to do some unpacking to get a sense of what it is supposed to be. But my goal in this appendix is to show that however one construes what they have in mind, there is no phenomenon, mechanism, or effect in the neighborhood of what the LSE group call a '*hawkmoth effect*' that poses a threat to prediction or projection that is in any way analogous to the butterfly effect. I will show that there are two ways in which the LSE group try to motivate a fear of the hawkmoth effect: the first by appealing to a family of mathematical results having to do with "structural stability," and the second by demonstrating an illustrative example, and then show that neither of these two paths is successful.

Whatever else it is, the hawkmoth effect is supposed to be a model-structure analog of the butterfly effect. In other words

Butterfly Effect:    Initial Conditions

::

Hawkmoth Effect:    Model Structure

The more technical term for the "butterfly effect," of course, is "sensitive dependence on initial conditions" (SDIC). We've given a good definition

---

[*] This appendix very closely follows portions of Navas, Nabergall and Winsberg (forthcoming). I thank Lukas Nabergall and Alejandro Navas for allowing me to reproduce some of the claims and arguments here. Interested readers should see the full paper for more details.
[1] (Smith 2002, Frigg et al. 2013, Frigg, Smith, and Stainforth 2013, Bradley et al. 2014, Frigg et al. 2014, Thompson 2013)
[2] Hereafter referred to as the "LSE group".

of the precise notion of SDIC that is most associated with the butterfly effect in Box 5.3.

> DEFINITION 1. **Exponential SDIC (for a map):** Let $\phi: X \to X$ be a time-evolution function on the manifold X. Say a dynamical system's behavior is *exponentially sensitive to initial conditions* if there exists $\lambda > 0$ such that for any state $x \in X$ and any *delta* $> 0$, almost all elements $y \in X$ satisfying $0 < d(x, y) < \delta$ are such that $d(\phi^t(x), \phi^t(y)) > e^{\lambda t} d(x, y)$.

That's if our system is specified with a map. If it is specified with a flow, then the definition becomes:

> DEFINITION 1*. **Exponential SDIC (for a flow):** Let $\phi: M \times \mathbb{R} \to M$ be a time-evolution function on the manifold X. Say a dynamical system's behavior is *exponentially sensitive to initial conditions* if there exists $\lambda > 0$ such that for any state $x \in X$ and any *delta* $> 0$, almost all elements $y \in X$ satisfying $0 < d(x, y) < \delta$ are such that $d(\phi(x,t), \phi(y,t)) > e^{\lambda t} d(x, y)$.

Intuitively, not only does exponential SDIC allow a system to get arbitrarily far away (subject to the boundedness of your dynamics) from where you would have gone by changing your initial state just a very small amount, but this "exponential" definition of SDIC (the most common notion) requires that you be able to get there very fast. More precisely, it says that there will be exponential growth in error – that every $1/\lambda$ units of time (called the "Lyapunov time") the distance between the trajectories picked out by the two close-by initial states will increase by a factor of $e$. Our definition also shares with most common definitions of SDIC that it is "strong." It requires that almost all the points near $x$ take you there, not just one. Strong, exponential SDIC is what people usually have in mind when they talk about the butterfly effect. Only when it has these two features does SDIC make "prediction practically impossible for more than short time periods."

A closely related property, as we saw in Boxes 5.4 and 5.5, is that of topological mixing (defined in Box 5.5). Topological mixing and SDIC are often taken to be the two necessary and sufficient conditions for "chaos." Informally, topological mixing (a crucial ingredient of chaos) occurs if, no matter how arbitrarily close I start, I will eventually be driven anywhere in the state space that I like. Here is the definition we gave in Chapter 5.

> DEFINITION 2. **Topological mixing (for a map)**[3]: A time-evolution function $\phi$ is called *topologically mixing* if for any pair of non-empty open sets $U$ and $V$, there exists a time $T > 0$ such that for all $t > T$, $\phi^t(U) \cap V \neq \varnothing$.

---

[3] The reader can easily write down a flow version of this definition.

It stands to reason, then, that the hawkmoth effect, if it exists, should involve a cluster of properties of dynamical systems that parallels those two, but where the notion of *two states being close* is replaced with the notion of *two equations of evolution* being close. It's easy to come up with an analog of topological mixing. Let's call it structural mixing.

> DEFINITION 3. **Structural mixing (for a family of flows)**: Let $\Phi$ be a space of time-evolution functions with metric $\delta$, $\phi \in \Phi$, $U$ be a non-empty open subset of $X$, and $\epsilon > 0$. Furthermore, for any time $t$, set $B_\epsilon(\phi)(t,x) = \{\phi'(t,x) \# \delta(\phi, \phi') < \epsilon\}$. We say that $\Phi$ is *structurally mixing at $\phi$* if for any state $x \in X$, there is a time $T$ such that for all $t > T$, $B_\epsilon(\phi)(t,x) \cap U \neq \varnothing$.

This definition uses a metric, $\delta$, to pick out the preferred topology on the space of evolution functions, but it makes more sense to relax it to an arbitrary topology, since there is not generally a natural metric on a space of evolution functions.

> DEFINITION 3*. **Structural mixing revamped**. Let $\Phi$ be a space of time-evolution functions, $\phi \in \Phi$, $U$ be a non-empty open subset of $X$, and $V$ be a non-empty open subset of $\Phi$ (in the appropriate topology) containing $\phi$. Furthermore, for any time $t$, set $V(t,x) = \{\phi'(t,x) \# \phi' \in V\}$. We say that $\Phi$ is *structurally mixing at $\phi$* if for any state $x \in X$, there is a time $T$ such that for all $t > T$, $V(t,x) \cap U \neq \varnothing$.

It's a bit harder to find a structural analog of SDIC. To see why, let's start by looking at a much weaker definition of SDIC than the one we discussed earlier – one that doesn't rely on exponential growth in error.

> DEFINITION 4. **SDIC to degree delta:** For a state space $X$ with metric $d$, say that the behavior of a dynamical system $(\mathbb{R}, X, \phi)$ with time-evolution function $\phi : \mathbb{R} \times X \to X$ is *sensitive to initial conditions to degree $\Delta$* if for every state $x \in X$ and every arbitrarily small distance $\delta > 0$, there exists a state y within distance $\delta$ of x and a time $t$ such that $d(\phi(t,x), \phi(t,y)) > \Delta$.

Definition 4 can be finessed into a structural analog. But now we absolutely need a metric of distance between two evolution-specifying functions. We are of course free to pick one, but it is worth keeping in mind that in many cases there will be no natural choice. That means that our definition will not provide a natural answer to the question "does this system display SDIC to degree delta?" But once we *have* picked a metric, we can define a notion of "sensitive dependence on model structure to degree $\Delta$" and define it roughly as follows.

DEFINITION 5. **Sensitive dependence on model structure to degree delta**: Let $\Phi$ be a space of time-evolution functions with metric $\delta$, $\phi \in \Phi$, and $\epsilon > 0$. We say that $\Phi$ *is sensitively dependent on model structure to degree* $\Delta$ at $\phi$ if for any state $x \in X$, there is a time $t$ such that for almost all $\phi' \in \Phi$ satisfying $\delta(\phi, \phi') < \epsilon$, we have $d(\phi'(t,x), \phi(t,x)) > \Delta$.

Definition 5 is already a bit strained since it relies on us picking a metric in a rather arbitrary way. But what we really want is a definition of *exponential* sensitive dependence on model structure. That is, we really want a structural analog of definition 1, not definition 4. A structural analog of definition 1 would really be a structural analog of the *butterfly effect*, and would generate analogous epistemological obstacles. But it is very hard to see how we could do this. We are already brushing under the rug the fact that definition 5 is assuming the existence of a metric of distance between evolution functions. But when we try to make an analog of definition 1, there is no single quantity that can grow in time in our new version. A structural analog of definition 1 would have to relate model distance with state space distance. It would have to coordinate a metric of model structure distance with a metric of state space distance. It's far less than obvious how this could avoid committing a category error.

## A.2 Two Routes to a Hawkmoth Effect

So much for what a hawkmoth effect should be like. What reason is there for thinking that there is such a thing? More precisely, what reason do these authors have for thinking that very many non-linear systems, including, importantly, the atmosphere-ocean-earth system, are best modeled by dynamical systems that exhibit it? The LSE group try to motivate the worry that the hawkmoth effect affects many such systems in two different ways.

The first argument comes in the form of an appeal to a variety of previously known mathematical results associated with the phenomenon of *structural stability*. And the second is their (in)famous demon's apprentice example. They usually insist that the first argument for the hawkmoth effect is the only argument they mean to offer, while the second is only meant as an illustrative example of something whose existence is established by the first argument. Still, we should look at each argument carefully. After all, even if the first argument fails to show that a large class of systems should be expected to be hawkmothish, the demon's apprentice example might show that hawkmoth behavior is still a danger to be reckoned with.

## A.3 The Structural Stability Route

The closest the LSE group come to giving a definition of the hawkmoth effect is in (Thompson 2013). Indeed, in what might be considered the flagship hawkmoth paper (Frigg et al 2014), what passes for a definition is: "Thompson (2013) introduced this term in analogy to the butterfly effect. The term also emphasizes that SME yields a worse epistemic position than SDIC: hawkmoths are better camouflaged and less photogenic than butterflies." (p. 39)

Looking in Thompson (2013) we find the following:

> In this chapter I introduce a result from the theory of dynamical systems and demonstrate its relevance for climate science. I name this result, for ease of reference, the Hawkmoth Effect (by analogy with the Butterfly Effect). (p. 211)

and

> The term "Butterfly Effect" has greatly aided communication and understanding of the consequences of dynamical instability of complex systems. It arises from the title of a talk given by Edward Lorenz in 1972: "Does the flap of a butterfly's wings in Brazil set off a tornado in Texas?".
>
> *I propose that the term "Hawkmoth Effect" should be used to refer to structural instability of complex systems.* The primary reason for proposing this term is to continue the Lepidoptera theme with a lesser-known but common member of the order. The Hawkmoth is also appropriately camouflaged, and less photogenic. (Thompson 2013, p. 213, emphasis added.)

Interestingly, the term "structural *in*stability" doesn't seem to appear very much in the mathematical literature. Nowhere is it described as producing an "effect." In fact, the discussion that we find in both (Thompson 2013, "5.2.2 Identifying structurally stable systems," p. 214–215), and even more so (Frigg et al. 2014, "4. From Example to Generalization," p. 45–47) are clearly influenced by (Pugh and Peixoto 2008) and though the word "stability" appears 65 times in (Pugh and Peixoto 2008), the word "instability" does not appear even once. Nor does the word "unstable." If you search for it, you can find the occasional article with "structural instability" in the title, but the results one finds in them are always discussed in terms of presence or absence of structural stability.

Why does this matter? It matters because structural stability does not have a complement with substantial features of its own, and the term "structural instability" suggests an overly close analogy to chaotic instability that no one in the mathematical literature ever had in mind. We can see why if we look closely at the notion of structural stability.

The first thing we should notice is that definitions of structural stability are definitions of ways of guaranteeing *to stay arbitrarily close*. And failure to stay arbitrarily close is not the same thing as being guaranteed to go arbitrarily far. But the butterfly effect does the latter, and not only the former.

Take an early definition of structural stability in two dimensions due to Andronov and Pontryagin as it is explained in Pugh and Peixoto (2008). They consider dynamical systems of the form

$$\frac{dx}{dt} = P(x, y), \quad \frac{dy}{dt} = Q(x, y)$$

defined on the disc $D^2$ in the $xy$-plane, with the vector field $(P, Q)$ entering transversally across the boundary $\partial D^2$.

> DEFINITION 6. **Roughness:** Let $p$ and $q$ be vector fields on $D^2$. We say that such a system is "*rough*," (as Andronov and Pontryagin put it) if, given $\epsilon > 0$, there exists $\delta > 0$ such that whenever $p(x, y)$ and $q(x, y)$, together with their first derivatives, are less than $\delta$ in absolute value, then the perturbed system
>
> $$\frac{dx}{dt} = P(x, y) + p(x, y), \quad \frac{dy}{dt} = Q(x, y) + q(x, y)$$
>
> is such that there exists an $\epsilon$-homeomorphism $h : D^2 \to D^2$ ($h$ moves each point in $D^2$ by less than $\epsilon$) which transforms the trajectories of the original system to the trajectories of the perturbed system.

Intuitively, this definition says of a particular evolution function that no matter how $\epsilon$-close to the trajectory of that evolution function I want to stay *for its entire history*, I can be guaranteed to find a $\delta$ such that all evolution functions within $\delta$ of my original one stay $\epsilon$-close to the original trajectory – where $\delta$ is a measure of how small both the perturbing function and their first derivatives are. This is an incredibly stringent requirement.

When an evolution function fails to obtain such a feature, therefore, it is as perverse to call it "structurally unstable" as it is to talk about a system being insensitively dependent on initial conditions, or to describe a function that fails to have a particular limit as "unlimited." It is a perversion that conflates the following two sorts of claims.

1) You can't be guaranteed to stay arbitrarily close by choosing an evolution function that is within some small neighborhood.

2) Small changes in the evolution function are sure to take you arbitrarily far away.

Absence of structural stability in the sense of "roughness" gives you the first thing, but nothing anywhere near approximating the second thing. But to claim that absence of structural stability is an analog of SDIC is to suggest that absence of structural stability gives you the second thing. That, moreover, is just the first difference. Notice that sensitive dependence on model structure, in order to be the analog of strong SDIC, (and to have a similar kind of epistemic bite) has to say "for almost any $\phi \in \Phi$". But roughness requires nothing of the sort. It only requires that, for each $\delta$, *one* trajectory in the entire set $\delta$-close trajectories be deformed by more than $\epsilon$.

And notice, by the way, that it would be impossible to even formulate a useful definition of structural instability that required that something like "almost all" the nearby models diverge. That's because there is no natural measure over the models. The state space of a classical system has an obvious measure: the Lebesgue measure. So it is easy to say things like "almost all the nearby states have such and such property." But spaces of diffeomorphisms have no such measure.[4]

This is what happens when you take the complement of a definition. Roughness is "strong." It requires that *not a single trajectory* diverges by more than $\epsilon$. But when you take the complement of a strong definition you get a very weak one. Here, the complement of structural stability only requires that, for each closeness threshold $\delta$, *one trajectory go astray by a tiny amount.*

---

[4] In reading some of their papers, and in conversation, one sometimes gets the impression that they think the result of Smale, which they summarize as follows "Smale, 1966, showed that structural stability is not generic in the class of diffeomorphisms on a manifold: the set of structurally stable systems is open but not dense," can stand in for a claim about the likelihood of a system being structurally stable. Or maybe of the likelihood of a nearby model being close if the first one is not structurally stable. This is the only thing we can find in the results they review about structural stability after they make the claim that "there are in fact mathematical considerations suggesting that the effects we describe are generic." (Frigg et al, 2014, p. 57) But first, climate models are not likely to be diffeomorphisms, so that's not the relevant universality class. Second, and more importantly, density is not a measure-theoretic notion, it is a topological one. A set can be dense and have measure zero (think of the rationals in the real number line – the rationals are of course not generic in the real line). There are even nowhere-dense sets that have arbitrarily high measure in the reals. There is a general point here: all the relevant notions associated with structural stability are topological, and they provide no information about likelihoods, or genericness. This is again because there is no natural measure on the space of equations of evolution.

To put the point simply, absence of structural stability in the roughness sense is inordinately weaker than either structural mixing or sensitive dependence on model structure. It's weaker in two ways: rather than requiring that *most* trajectories (indeed as we have seen there is no coherent notion of "most" here) go *far* away, it only requires that *one* trajectory go *more than a very small epsilon* away. And hence it is much weaker than the LSE group or the hawkmoth analogy suggest.

We still haven't talked, moreover, about the exponential definition of SDIC and the idea of exponential error growth, which is so fundamental to the epistemological impact of chaos. To get a structural analog of this, we would need to formulate a definition that captured the idea that the small error in *model space* could grow very fast is the *space of states*. But this looks unlikely for the case of failure of roughness. The reason is that in their definition the metric of model distance does not live in the same space as where the metric of state space distance lives. It would be strange and confusing to relate these two metrics in a single equation.

And things get even worse if we move from what Pugh and Peixoto call the "pre-history" of structural stability to its modern formulation. In the modern formulation, no metric is specified – neither between two different diffeomorphisms, nor between two trajectories.

The modern formulation of structural stability goes as follows:

DEFINITION 7. **Structural stability of a map:** If $D$ is the set of self-diffeomorphisms of a compact smooth manifold $M$, and $D$ is equipped with the $C^1$ topology then $f \in D$ is *structurally stable* if and only if for each $g$ in some neighborhood of $f$ in $D$ there is a homeomorphism $h : M \to M$ such that $f$ is topologically conjugate to each nearby $g$. In other words, that

$$
\begin{array}{ccc}
M & \xrightarrow{\ f\ } & M \\
h \downarrow & & \downarrow h \\
M & \xrightarrow{\ g\ } & M
\end{array}
$$

commutes, that is, $h(f(x)) = g(h(x))$ for all $x \in M$.

This definition makes it clear that the most general formulation of it is not metrical at all. It is topological. It says nothing at all about diffeomorphisms that are "a small distance away." It talks about diffeomorphisms that are in topological neighborhoods of each other. These are only very loosely related. And it doesn't talk at all about trajectories taking you some distance away. It talks about there failing to be a homeomorphism (a topology preserving transformation that cares nothing about distances) between the two trajectories. Using the analogy of a rubber sheet that is often used to explain topological notions, it roughly says, intuitively, that if you replace $f$ by any of the diffeomorphisms $g$ in some neighborhood of $f$ then the new entire state-space diagram of $g$ will be one that could have been made just by stretching or unstretching (by deforming it in the way one can deform a rubber sheet without tearing it) the state-space diagram of $f$. And so in fact, in some cases, structural stability tells you nothing at all, let alone anything strong enough, about how far away a slightly perturbed model will take you. It simply supplies no metrical information whatsoever.

So let's review what it amounts to for a system defined by a diffeomorphic map $f$, applied to a manifold, to fail to be structurally stable. It means that if you look at the space of diffeomorphisms around $f$, you will be unable to find a neighborhood around $f$ that is guaranteed not to contain a single other diffeomorphism $g$ that is qualitatively different than $f$ in essentially the following sense: that you cannot smoothly deform $f$ into $g$. The Father Guido Sarducci version of what we have learned so far is this: *You don't get structural "instability" just by replacing "small changes in initial conditions" in SDIC with "small changes in model structure" because, both with regard to how much error you can get, and with respect to how many nearby trajectories will do it, SDIC says things are maximally bad, while structural instability merely says they will not be maximally good. And SDIC includes metrical claims, while its incoherent for structural stability to be given a metrical form.*

Let us put this another way. Absence of structural stability is an incredibly weak condition on three dimensions: it need not take you far (the relevant notion is defined in the absence of any metric at all), it need not take you there at all fast, and there only needs to be one model in your entire neighborhood that does it. So absence of structural stability has no interesting consequences for the predictive power of a model, even in the presence of model structure uncertainty – so long as you are not interested in infinitely long predictions (the second dimension) that are topologically exact (the first dimension) and underwritten by mathematical certainty (the third dimension), rather than, say, overwhelming probability.[5]

---

[5] This is all of course because structural stability was a notion developed by people interested in achieving the mathematical certainty of proof while using perturbations – they were not interested in finding predictive accuracy. After all, they were studying the solar system. No one was worried that, until they found stability results for the solar system, its dynamical study would be embroiled in confusion or maladapted to quantitative prediction.

## A.4 The Demon's Apprentice Route

The other argument for a hawkmoth effect is an argument by example: in Frigg et al. (2014), the LSE group postulate the existence of a demon that is omniscient regarding the exact initial conditions of a given system, the true dynamical model of the system, and the computational output of such a model at any future time, for any initial conditions, to arbitrary precision. Such a demon also has two apprentices: a senior apprentice, who has omniscience of computation and dynamics yet lacks that of initial conditions, and a freshman apprentice, who has computational omniscience but not that of model structure nor initial conditions.

The problems of the senior apprentice can be overcome by Probabilistic Initial Condition Ensemble Forecasting (PICEF). In PICEF, instead of using a single point in the state space as the initial conditions for the dynamical system, we substitute in a probability distribution over the entire state space. In this way, the practical limitations of initial condition uncertainty can be mitigated: a point prediction for the state of the system in the future is replaced by a distribution over possible future states which may still inform practical considerations.

According to the LSE group, however, there is no such solution for the freshman's ignorance. Aside from the initial condition uncertainties which reduce the precision of a system's trajectories, the freshman apprentice is beset by unreliability in the very dynamics by which initial states are evolved. This unreliability, they claim, is not easily resolved nor easily dismissed. And its consequences, they hold, are severe.

To illustrate this severity, et al., consider the logistic map, defined below (and in Chapter 5) with equation A.1, and a "similar" equation that represents the true model of some physical system, equation A.2. They show that, given enough time, the two equations evolve the same distribution of initial conditions to very different regions of state space. We've already seen that the absence of structural stability is not, in general, as severe as the results of the demon's apprentice example suggest. Absence of structural stability does not generally lead to wide divergence, nor does its absence imply anything about the majority of nearby models. It only takes one stray model in the neighborhood to violate the definition. But we can still ask whether the demon example is at least *a* possible illustration of the absence of structural stability. And we can still ask if it provides a worthwhile cautionary tale of its own. The answers to both of these questions, alas, is "no."

Why is the demon example not a possible illustration of the absence of structural stability? To see why, we need to look at the logistic equation, which is part of the demon example:

$$x_{t+1} = 4x_t\left(1 - x_t\right).\qquad\qquad\text{(A.1)}$$

The logistic equation, as we have seen, is a map. Structural stability is defined for both maps and flows. Definition 7 was for maps, and the definition for flows is:

> DEFINITION 8. **Structural stability for a flow**: If $X$ is the set of smooth vector fields on a manifold M equipped with the $C^1$ topology, then the flow generated by $x \in X$ is *structurally stable* if and only if for each $y$ in the neighborhood of $x$ in $X$ there is a homeomorphism $h : M \to M$ that sends the orbits of $x$ to the orbits of $y$ while preserving the orientation of the orbits.

But notice that definition 7 (for maps) does not apply to any old map, and the relevant family, (the "universality class"), of maps that count as "in the neighborhood" is *not all maps*. The definition only applies if the map is a diffeomorphism and the universality class is also diffeomorphisms. To be a diffeomorphism, a map has to have some added conditions:

1) The map has to be differentiable.
2) The map has to be a bijection (its inverse must also be a function).
3) The inverse of the map has to be differentiable.

The obvious problem is that the logistic map is not a bijection![6] Every number other than 1 has two preimages. For example, both 0.3 and 0.7 map to 0.84. So 0.84 has no unique preimage and there is no function that is the inverse of the logistic equation. But this means that the logistic map isn't even the right category of object to be structurally stable or not.[7]

---

[6] And neither is equation A.3 (its "nearby" neighbor).

[7] There are other nearby puzzles about what the LSE group could possibly have thought they were on about: the best model climate science could write down – that is the real true, partial-differential-equation-specified, undiscretized model – would have the form $\phi: M \times \mathbb{R} \to M$. It might or might not meet the additional criteria for being a flow, but it is certainly not of the form $\phi: M \to M$, which is the general form of a map. Once we start to think about a discretized model, however, the model does take the form $\phi: M \to M$. Even if we had the perfect climate model and it *were* a flow, a discretization of it (in time) would necessarily have the form of a map. And no map is in the right universality class – for purposes of structural stability – of a flow. In the sense relevant to structural stability, its simply a category error to ask if a discretization of a dynamical system of the form $\phi: M \times \mathbb{R} \to M$ is "nearby to" the undiscretized system. Climate models run on computers are all imperfect, but they don't live in the same universe of functions as the "true" model of the climate does.

Of course we are free to make up our own definition of structural stability that applies to all maps, and where the universality class is all maps. But if we do, if we expand our model class beyond the space of diffeomorphisms to the space of all maps, then *the very notion of structural stability becomes empty*. You simply won't find many structurally stable maps on this definition. Consider the simplest map there is:

$$x_i = C \quad \text{for all } i \text{ and some constant } C. \tag{A.2}$$

This map is simple but, of course, not a bijection. Yet it also would not come out as "structurally stable" on our new definition. According to the definition we require all $g$ in some neighborhood to satisfy $h(f(x)) = g(h(x))$. But if $f$ is constant, $h(f(x))$ is constant, so $g(h(x))$ has to be constant for the definition to hold. But that condition is easy to violate with many of the $g$'s in any neighborhood of $f$ if $g$ can be any map. But this means that the logistic map is no more "structurally unstable" than the function given by equation A.2 is. Which means whatever the demon example illustrates, it actually has nothing at all to do with non-linearity. Technically, of course, equation A.2 isn't linear. But its also not exactly what people have in mind when they think of non-linearity! And of course, once you open up the model class to non-diffeomorphisms, $f(x) = 3x$ (an obviously linear map) will also be structurally unstable.

So when Frigg et al. (2014) write, "The relation between structural stability and the Demon scenario is obvious: if the original system is the true dynamics, then the true dynamics has to be structurally stable for the Freshman's close-by model to yield close-by results,"(p. 47) they are saying something very misleading. In fact, the relation between structural stability and the demon is at best murky – because the logistic equation is not even a candidate for structural stability. And as we have seen, it is simply false that a model has to be structurally stable for a nearby model to produce nearby results. That is straightforwardly a misreading of the definition. And it is straightforwardly misleading to suggest that non-linearity is the culprit. If you open up the definition to include arbitrary maps, all kinds of incredibly simple maps (linear or otherwise) become "structurally unstable." If you think absence of structural stability is the hawkmoth effect, and if you think the logistic equation (despite not being a diffeomorphism) displays the hawkmoth effect, then necessarily you will have to say that $x_i = C$ displays the hawkmoth effect too. This is not a happy outcome!

Okay, but still, even if the logistic map is not a candidate for structural stability, surely the demon example still shows that two very nearby models can lead to radically different PICEF predictions, right? We saw in section A.3 that structural stability and its absence did not underwrite an analog of the butterfly effect. But maybe the demon example does by itself. This, after all, we can see with our own eyes in the Frigg et al (2014) paper. Not so fast. Let's look carefully at the two models in the example: the freshman apprentice's model and the demon's model.

$$x_{t+1} = 4x_t(1 - x_t) \tag{A.1}$$

$$\tilde{x}_{t+1} = (1 - \epsilon)4\tilde{x}_t(1 - \tilde{x}_t) + \frac{16\epsilon}{5}\left[\tilde{x}_t(1 - 2\tilde{x}_t^2 + \tilde{x}_t^3)\right] \tag{A.3}$$

Equation A.3 is the demon's and senior apprentice's model, the "true" model in this scenario, and equation A.1 is the freshman apprentice's model, the "approximate" model.

On first glance, these equations do not look very similar. But the LSE group argue that they are in fact similar. They argue this by arguing that the appropriate metric of similarity should be based on an output comparison of the two models over one time step: Call the maximum difference that the two models can produce for any arbitrary input $\epsilon$, the maximum one-step error. If $\epsilon$ is sufficiently small, then the two models can be said, according to the LSE group, to be very similar. The most important point to make here is that this *is much too weak of a condition* for considering two models similar. It is not for nothing, moreover, that the modern literature on structural stability is topological and not metrical. There just isn't anything sufficiently general and sufficiently natural to say about how to measure the distance between two models, two diffeomorphisms, or two flows. The model of equation A.1 and the model of equation A.3 are not appropriately similar for drawing the conclusions that the LSE group draw. The best way to see why maximum one-step error is not a good measure of model similarity is to consider a theorem that Nabergall, Navas and I proved in our (forthcoming).

## A.5  Small Arbitrary Model Perturbations

Suppose we are given a difference equation of the form

$$x_{n+1} = f(x_n) \tag{A.4}$$

where $x_i \in \mathbb{R}$ and $f : A \rightarrow B$ is an arbitrary function from the bounded interval $A \subset \mathbb{R}$ into the bounded interval $B \subset \mathbb{R}$. Note that the logistic map takes this form with $f(x_n) = 4x_n(1 - x_n)$. Then we have the following result:

THEOREM 1: *Given any function* $g : A \rightarrow B$ *and* $\epsilon > 0$, *there exists* $\delta > 0$ and $\eta > 0$ *such that the maximum one-step error of*

$$x'_{n+1} = \eta f(x'_n) + \delta g(x'_n) \qquad \text{(A.5)}$$

from equation A.4 is at most $\epsilon$ and $x'_{n+1} \in B$.

Observe that equation A.4 takes the form of equation A.5 with $f(x'_n) = 4x'_n(1 - x'_n)$, $g(x'_n) = (16/5)[x'_n(1 - 2x'^2_n + x'^3_n)]$, $\eta = 1 - \epsilon$, and $\delta = \epsilon$.

There are at least two ways in which this result undermines the claim that the demon's apprentice example demonstrates the existence of a hawk-moth effect which is an epistemological analog of the butterfly effect.

The existence of small arbitrary model perturbations demonstrated in the above theorem for first-order difference equations, of which the logistic map is an example, shows that the perturbation presented in the demon example is only one possible perturbation amongst the infinite space of admissible perturbations that are close to the logistic map under the max-imum one-step error model metric. In fact, as the argument demonstrates, we can perturb our model in any way we wish and still remain as close to the initial model as desired. It should therefore be no surprise that we can find models close to the logistic map that generate trajectories in the state space vastly diverging from the logistic map over certain time intervals; indeed, we should expect to find nearby models that exhibit essentially any behavior we want, including some which vastly deviate from the logistic map across any given time interval and others which remain arbi-trarily close to the logistic map for all future times. In particular, there is no a priori reason we should expect that the modified logistic map is an example of a commonly occurring small model error. The butterfly effect is so important because numerically small differences between the true value of a system variable and its measured value are absolutely common and normal. But what reason is there to think that climate scientists make mistakes about the order of the polynomials their models should have? Or that they sometimes write down exponential functions when they should have written down sinusoidal ones? Corollarily, what reason is there to think that small model errors of the kind we would expect to find in cli-mate science, atmospheric science, and other domains of non-linear mod-eling will normally produce deviations on such short timescales as they do in the demon example? Why would we let such a weirdly concocted example do any "burden of proof shifting" of the kind the LSE group demand of us? Consequently, at the very least, the demon example only retains its relevance as evidence for the epistemic force of the so-called hawkmoth effect if it is accompanied by a strong argument showing how it is precisely this sort of perturbation which is often encountered when constructing weather-forecasting models. But there may be good reasons to think this is not the case.

More generally, Theorem 1 indicates that the maximum one-step error metric is quite simply too easily satisfied and does not really get at what makes two models similar or close. It would be difficult indeed to argue that the difference equations

$$x_{n+1} = 2x_n \quad \text{and} \quad x_{n+1} = 2x_n + .004e^{x_n} - .03\sin x_n \qquad \text{(A.6)}$$

are highly similar and ought to be considered "close" in model space simply because after one time step they do not yet produce significantly diverging trajectories in state space – the newly added sinusoidal and exponential terms behave so differently from the linear term present in the original equation that we would certainly not want to call these two models "close." Furthermore, we would in particular definitely not expect to be able to predict well the long-term behavior of a physical system using both of these equations since the perturbations introduced in the second equation model entirely different physical dynamics.

In sum, absence of structural stability is not a hawkmoth effect, because it is much too weak of an "effect." And the demon example, while it displays what looks like a strong "effect," is not a case of structural instability. It is also not a good illustration of a hawkmoth effect both because it is not an interesting case of two "nearby" models and because it draws on much too liberal of a universality class.

# References

## Special Note Regarding Citations to IPCC Assessment Reports

In making citations to the IPCC's assessment reports, rather than using the cumbersome and confusing practice of the (Author Date) system, all of my citations are of the form IPCC, AR5, WG1, 11.1, p. 959 (unless a page number is not appropriate). This example refers to Chapter 11 of Working Group 1's contribution to the Fifth Assessment Report (AR5), section 11.1, page 959. This makes finding the relevant section on the IPCC's webpage much easier.

The three IPCC assessment reports that are cited in this volume are:

1) IPCC *Third Assessment Report: Climate Change 2001 (TAR)*
2) IPCC *Fourth Assessment Report: Climate Change 2007 (AR4)*
3) IPCC *Fifth Assessment Report (AR5)*

Almost all citations to any of the above are to the reports of "Working Group 1" which refer, respectively to

1) Working Group I: The Scientific Basis (for TAR)
2) Working Group I Report: "The Physical Science Basis" (for AR4)
3) Working Group I Report: "Climate Change 2013: The Physical Science Basis" (for AR5)

Alexander, Kaitlin and Stephen M. Easterbrook. 2015. "The Software Architecture of Climate Models: A Graphical Comparison of CMIP5 and EMICAR5 Configurations." *Geoscientific Model Development* 8 (4): 1221–1232.

Allison, I., N. L. Bindoff, R. A. Bindschadler, P. M. Cox, N. De Noblet, M. H. England, J. E. Francis et al. 2009. "The Copenhagen Diagnosis." *The University of New South Wales Climate Change Research Centre*, Sydney: 1-60. www.researchgate.net/profile/E_Rignot/publication/51997579_The_Copenhagen_Diagnosis/links/00b7d5159bf09dd86b000000.pdf.

Barnes, Eric Christian. 2008. *The Paradox of Predictivism*. 1st edition. Cambridge; New York: Cambridge University Press.

Biddle, Justin B. and Anna Leuschner. 2015. "Climate Skepticism and the Manufacture of Doubt: Can Dissent in Science Be Epistemically Detrimental?" *European Journal for Philosophy of Science* 5 (3): 261–278.

Biddle, Justin and Eric Winsberg. 2009. "Value Judgements and the Estimation of Uncertainty in Climate Modeling." In Magnus, P.D. and Busch, J., eds. *New Waves in Philosophy of Science*, p. 172–197. London: Palgrave McMillan.

Bogen, James and James Woodward. 1988. "Saving the Phenomena." *The Philosophical Review* 97 (3): 303–352.

Bovens, Luc and Stephan Hartmann. 2003. *Bayesian Epistemology*. Oxford University Press.

Bradley, Richard and Katie Steele. 2015. "Making Climate Decisions." *Philosophy Compass* 10 (11): 799–810.

Bradley, Seamus, Roman Frigg, Hailiang Du and Leonard A. Smith. 2014. "Model Error and Ensemble Forecasting: A Cautionary Tale." *Scientific Explanation and Methodology of Science* 1: 58–66.

Bray, Dennis. 2010. "The Scientific Consensus of Climate Change Revisited." *Environmental Science & Policy* 13 (5): 340–350.

Bray, Dennis and Hans von Storch. 2010. *CliSci2008: A Survey of the Perspectives of Climate Scientists Concerning Climate Science and Climate Change*. GKSS-Forschungszentrum Geesthacht Geesthacht. www.hvonstorch.de/klima/pdf/GKSS_2010_9.CLISCI.pdf.

Brook, Barry. 2008. "So Just Who Does Climate Science?" Brave New Climate. August 31. https://bravenewclimate.com/2008/08/31/so-just-who-does-climate-science/.

Broome, John. 2012. *Climate Matters: Ethics in a Warming World*. 1st edition. New York: W. W. Norton & Company.

Bump, Philip. 2015. "Jim Inhofe's Snowball Has Disproven Climate Change Once and for All." *The Washington Post*, February 26, sec. The Fix. www.washingtonpost.com/news/the-fix/wp/2015/02/26/jim-inhofes-snowball-has-disproven-climate-change-once-and-for-all/.

Calcott, Brett. 2011. "Wimsatt and the Robustness Family: Review of Wimsatt's Re-Engineering Philosophy for Limited Beings." *Biology & Philosophy* 26 (2): 281–293.

Carlton, J. S., Rebecca Perry-Hill, Matthew Huber and Linda S. Prokopy. 2015. "The Climate Change Consensus Extends beyond Climate Scientists." *Environmental Research Letters* 10 (9): 094025.

Chaboureau, Jean-Pierre and Peter Bechtold. 2002. "A Simple Cloud Parameterization Derived from Cloud Resolving Model Data: Diagnostic and Prognostic Applications." *Journal of the Atmospheric Sciences* 59 (15): 2362–2372. doi:10.1175/1520-0469(2002)059<2362:ASCPDF>2.0.CO;2.

Churchman, C. West. 1948. "Statistics, Pragmatics, Induction." *Philosophy of Science* 15 (3): 249–268.

1956. "Science and Decision Making." *Philosophy of Science* 23 (3): 247–249.

Clark, Andy. 1987. "The Kludge in the Machine." *Mind & Language* 2 (4): 277–300.

1998. *Being There: Putting Brain, Body, and World Together Again*. Cambridge, MA: MIT Press.

Cohen, L. Jonathan. 1995. *An Essay on Belief and Acceptance*. Reprint edition. Oxford: Clarendon Press.

Cook, John, Naomi Oreskes, Peter T. Doran, William R.L. Anderegg, Bart Verheggen, Ed W. Maibach, J. Stuart Carlton et al. 2016. "Consensus on Consensus: A Synthesis of Consensus Estimates on Human-Caused Global Warming." *Environmental Research Letters* 11 (4): 048002.

Daron, Joseph D. and David A. Stainforth. 2013. "On Predicting Climate under Climate Change." *Environmental Research Letters* 8 (3): 034021.

Doran, Peter T. and Maggie Kendall Zimmerman. 2009. "Examining the Scientific Consensus on Climate Change." *Eos, Transactions American Geophysical Union* 90 (3): 22–23.

Douglas, Heather. 2000. "Inductive Risk and Values in Science." *Philosophy of Science* 67 (4): 559–579.

    2005. "Inserting the Public into Science." In *Democratization of Expertise?*, 153–169. Springer. http://link.springer.com/content/pdf/10.1007/1-4020-3754-6_9.pdf.

    2009. *Science, Policy, and the Value-Free Ideal*. University of Pittsburgh Press.

Dunne, John. 2006. "Towards Earth System Modelling: Bringing GFDL to life," in "'The Australian Community Climate and Earth System Simulator (ACCESS) - challenges and opportunities': extended abstracts of presentations at the eighteenth annual BMRC Modelling Workshop 28 November - 1 December 2006." A.J. Hollis and A.P. Kariko (editors) downloaded from www.cawcr.gov.au/publications/BMRC_archive/researchreports/RR123.pdf, accessed August 12, 2014.

"Earth System Modeling Framework." 2016. Wikipedia. https://en.wikipedia.org/w/index.php?title=Earth_System_Modeling_Framework&oldid=749938673.

Elliott, Kevin C. 2011. "Direct and Indirect Roles for Values in Science." *Philosophy of Science* 78 (2): 303–324.

Elliott, Kevin C. and Daniel J. McKaughan. 2014. "Nonepistemic Values and the Multiple Goals of Science." *Philosophy of Science* 81 (1): 1–21.

Elliott, Kevin C. and David B. Resnik. 2014. "Science, Policy, and the Transparency of Values." *Environmental Health Perspectives (Online)* 122 (7): 647.

Farnsworth, Stephen J. and S. Robert Lichter. 2012. "The Structure of Scientific Opinion on Climate Change." *International Journal of Public Opinion Research* 24 (1): 93–103.

Fehr, Carla and Kathryn S. Plaisance. 2010. "Socially Relevant Philosophy of Science: An Introduction." *Synthese* 177 (3): 301–316.

Fraassen, Bas C. van. 2010. *Scientific Representation: Paradoxes of Perspective*. Reprint edition. Oxford; New York: Oxford University Press.

Frigg, Roman, Seamus Bradley, Hailiang Du and Leonard A. Smith. 2014. "Laplace's Demon and the Adventures of His Apprentices." *Philosophy of Science* 81 (1): 31–59.

Frigg, Roman, Seamus Bradley, Reason L. Machete and Leonard A. Smith. 2013. "Probabilistic Forecasting: Why Model Imperfection Is a Poison Pill." In

*New Challenges to Philosophy of Science*, 479–491. Springer. http://link.springer.com/chapter/10.1007/978-94-007-5845-2_39.

Frigg, Roman and Stephan Hartmann. 2012. "Models in Science." *Stanford Encyclopedia of Philosophy*. http://plato.stanford.edu/archives/fall2012/entries/models-science.

Frigg, Roman and Julian Reiss. 2009. "The Philosophy of Simulation: Hot New Issues or Same Old Stew?" *Synthese* 169 (3): 593–613.

Frigg, Roman, Leonard A. Smith and David A. Stainforth. 2013. "The Myopia of Imperfect Climate Models: The Case of UKCP09." *Philosophy of Science* 80 (5): 886–897.

2015. "An Assessment of the Foundational Assumptions in High-Resolution Climate Projections: The Case of UKCP09." *Synthese* 192 (12): 3979–4008.

Frisch, Mathias. 2013. "Modeling Climate Policies: A Critical Look at Integrated Assessment Models." *Philosophy & Technology* 26 (2): 117–137. Reprinted in *Climate Modeling: Philosophical and Conceptual Issues,* edited by Elisabeth A. Lloyd and Eric Winsberg. 2018. Palgrave McMillan.

2015. "Predictivism and Old Evidence: A Critical Look at Climate Model Tuning." *European Journal for Philosophy of Science* 5 (2): 171–190.

2017. "Climate Policy in the Age of Trump." *Kennedy Institute of Ethics Journal* 27 (S2): 87–106.

Gleckler, Peter J., Karl E. Taylor and Charles Doutriaux. 2008. "Performance Metrics for Climate Models." *Journal of Geophysical Research: Atmospheres* 113 (D6). http://onlinelibrary.wiley.com/doi/10.1029/2007JD008972/full.

Golaz, Jean-Christophe, Larry W. Horowitz and Hiram Levy. 2013. "Cloud Tuning in a Coupled Climate Model: Impact on 20th Century Warming." *Geophysical Research Letters* 40 (10): 2246–2251.

Goldman, Alvin and Thomas Blanchard. 2016. "Social Epistemology." In *The Stanford Encyclopedia of Philosophy*, edited by Edward N. Zalta, Winter 2016. Metaphysics Research Lab, Stanford University. https://plato.stanford.edu/archives/win2016/entries/epistemology-social/.

Goldman, Alvin I. 2001. "Experts: Which Ones Should You Trust?" *Philosophy and Phenomenological Research* 63 (1): 85–110.

Goodwin, William M. 2015. "Global Climate Modeling as Applied Science." *Journal for General Philosophy of Science* 46 (2): 339–350.

Gordon, Neil D. and Stephen A. Klein. 2014. "Low-Cloud Optical Depth Feedback in Climate Models." *Journal of Geophysical Research: Atmospheres* 119 (10): 2013JD021052. doi:10.1002/2013JD021052.

Grier, David. 2005. *When Computers Were Human.* Princeton University Press. http://press.princeton.edu/titles/7999.html.

Hahn, Hans, Otto Neurath and Rudolf Carnap. 1929. *The Scientific Conception of the World: The Vienna Circle.*

Hájek, Alan. 2003. "Interpretations of Probability." *The Stanford Encyclopedia of Philosophy*. Citeseer. http://citeseerx.ist.psu.edu/viewdoc/summary?doi=10.1.1.125.8314.

Held, Isaac and David Randall. 2011. "Point/Counterpoint." *IEEE Software* 28 (6): 62–65.

Hoefer, Carl. 2007. "The Third Way on Objective Probability: A Sceptic's Guide to Objective Chance." *Mind* 116 (463): 549–596.

Hollingsworth, Barbara. 2015. "97% Climbs to 99.5%: Obama Increases Percentage of Scientists Who Agree on Climate Change." CNS News, December 2. www.cnsnews.com/news/article/barbara-hollingsworth/obama-ups-climate-change-consensus-paris-995-scientists.

Howard, Don A. 2003. "Two Left Turns Make a Right: On the Curious Political Career of North American Philosophy of Science at Midcentury." In Alan W. Richardson and Gary L. Hardcastle, eds. *Logical Empiricism in North America*, pp. 25–93. Minneapolis, MN: University of Minnesota Press.

Huebner, Bryce, Rebecca Kukla and Eric Winsberg. 2017. "Making an Author in Radically Collaborative Research." In Thomas Boyer-Kassem, Conor Mayo-Wilson and Michael Weisberg, eds. *Scientific Collaboration and Collective Knowledge*, 95–116. Oxford University Press.

Hume, David. 2011. *An Enquiry Concerning Human Understanding*. Hollywood, FL: Simon & Brown.

Intemann, Kristen. 2015. "Distinguishing between Legitimate and Illegitimate Values in Climate Modeling." *European Journal for Philosophy of Science* 5 (2): 217–232.

2017. "Who Needs Consensus Anyway? Addressing Manufactured Doubt and Increasing Public Trust in Climate Science." *Public Affairs Quarterly* 31 (3): 189–208.

Jeffrey, Richard C. 1956. "Valuation and Acceptance of Scientific Hypotheses." *Philosophy of Science* 23 (3): 237–246.

Kellert, Stephen H. 1993. *In the Wake of Chaos: Unpredictable Order in Dynamical Systems*. 1st edition. Chicago: University Of Chicago Press.

Kitcher, Phillip. 1993. *The Advancement of Science: Science without Legend, Objectivity without Illusions*. Oxford University Press.

Klein, Stephen A. and Alex Hall. 2015. "Emergent Constraints for Cloud Feedbacks." *Current Climate Change Reports* 1 (4): 276–287.

Knight, Frank H. 2012. *Risk, Uncertainty and Profit*. North Chelmsford, MA: Courier Corporation.

Knutti, Reto. 2018. "Climate Model Confirmation: From Philosophy to Predicting Climate in the Real World." In *Climate Modeling: Philosophical and Conceptual Issues*, edited by Elisabeth A. Lloyd and Eric Winsberg. London: Palgrave McMillan.

Knutti, Reto and Gabriele C. Hegerl. 2008. "The Equilibrium Sensitivity of the Earth's Temperature to Radiation Changes." *Nature Geoscience* 1 (11): 735–743.

Kourany, Janet A. 2003a. "A Philosophy of Science for the Twenty-First Century." *Philosophy of Science* 70 (1): 1–14.

2003b. "Reply to Giere." *Philosophy of Science* 70 (1): 22–26.

Kuhn, Thomas S. 2012. *The Structure of Scientific Revolutions*. University of Chicago Press.

Küppers, Günter and Johannes Lenhard. 2006. "From Hierarchical to Network-like Integration: A Revolution of Modeling Style in Computer-Simulation." In *Simulation*, 89–106. Springer. http://link.springer.com/content/pdf/10.1007/1-4020-5375-4_6.pdf.

Lenhard, Johannes and Eric Winsberg. 2010. "Holism, Entrenchment, and the Future of Climate Model Pluralism." *Studies in History and Philosophy of Science Part B: Studies in History and Philosophy of Modern Physics* 41 (3): 253–262.

Levins, Richard. 1966. "The Strategy of Model Building in Population Biology." *American Scientist* 54 (4): 421–431.

Lloyd, Elisabeth A. 2009. "I – Elisabeth A. Lloyd: Varieties of Support and Confirmation of Climate Models." In *Aristotelian Society Supplementary Volume* 83: 213–232. Oxford University Press. http://aristoteliansupp.oxfordjournals.org/content/83/1/213.abstract.

2010. "Confirmation and Robustness of Climate Models." *Philosophy of Science* 77 (5): 971–984.

2012. "The Role of 'Complex' Empiricism in the Debates about Satellite Data and Climate Models." *Studies in History and Philosophy of Science Part A* 43 (2): 390–401. Reprinted in *Climate Modeling: Philosophical and Conceptual Issues,* edited by Elisabeth A. Lloyd and Eric Winsberg. 2018. Palgrave McMillan.

2015. "Model Robustness as a Confirmatory Virtue: The Case of Climate Science." *Studies in History and Philosophy of Science Part A* 49: 58–68.

Lloyd, Elisabeth A. and Eric Winsberg, eds. 2018. *Climate Modeling: Philosophical and Conceptual Issues.* London: Palgrave MacMillan.

Longino, Helen E. 1990. *Science as Social Knowledge: Values and Objectivity in Scientific Inquiry.* Princeton University Press.

1996. "Cognitive and Non-Cognitive Values in Science: Rethinking the Dichotomy." In *Feminism, Science, and the Philosophy of Science*, 39–58. Springer. http://link.springer.com/chapter/10.1007/978-94-009-1742-2_3.

2002. *The Fate of Knowledge.* Princeton University Press.

Machamer, Peter and Heather Douglas. 1999. "Cognitive and Social Values." *Science & Education* 8 (1): 45–54.

Mastrandrea, Michael D., Christopher B. Field, Thomas F. Stocker, Ottmar Edenhofer, Kristie L. Ebi, David J. Frame, Hermann Held et al. 2010. "Guidance Note for Lead Authors of the IPCC Fifth Assessment Report on Consistent Treatment of Uncertainties." http://pubman.mpdl.mpg.de/pubman/item/escidoc:2147184/component/escidoc:2147185/uncertainty-guidance-note.pdf.

Mauritsen, Thorsten, Bjorn Stevens, Erich Roeckner, Traute Crueger, Monika Esch, Marco Giorgetta, Helmuth Haak et al. 2012. "Tuning the Climate of a Global Model." *Journal of Advances in Modeling Earth Systems* 4 (3). http://onlinelibrary.wiley.com/doi/10.1029/2012MS000154/full.

McMullin, Ernan. 1982. "Values in Science." In *PSA: Proceedings of the Biennial Meeting of the Philosophy of Science Association*, 1982: 3–28. Philosophy of

Science Association. www.journals.uchicago.edu/doi/abs/10.1086/psaprocbi enmeetp.1982.2.192409.

Mearns, Linda O., Melissa S. Bukovsky, Sarah C. Pryor and Victor Magaña. 2014. "Downscaling of Climate Information." In *Climate Change in North America*, 201–250. Springer. http://link.springer.com/chapter/10.1007/ 978-3-319-03768-4_5. Reprinted in *Climate Modeling: Philosophical and Conceptual Issues,* edited by Elisabeth A. Lloyd and Eric Winsberg. 2018. Palgrave McMillan.

Miller, Boaz. 2013. "When Is Consensus Knowledge Based? Distinguishing Shared Knowledge from Mere Agreement." *Synthese* 190: 1293–1316.

Morgan, Mary S. and Margaret Morrison, eds. 1999. *Models as Mediators: Perspectives on Natural and Social Science.* N edition. Cambridge; New York: Cambridge University Press.

Navas, Alejandro, Lukas Nabergall and Eric Winsberg. Forthcoming. "On the Proper Classification of Lepidoptera: Does the Absence of Structural Stability Produce a 'Hawkmoth Effect' That Undermines Climate Modeling?"

Neurath, Marie and Robert S. Cohen, eds. 1973. *Otto Neurath: Empiricism and Sociology.* 1973 edition. Dordrecht: D. Reidel Publishing Company.

Nordhaus, W. 2008. *A Question of Balance: Weighing the Options on Global Warming Policies.* New Haven; London: Yale University Press.

Nuccitelli, Dana, Robert Way, Rob Painting, John Church and John Cook. 2012. "Comment on 'Ocean Heat Content and Earth's Radiation Imbalance. II. Relation to Climate Shifts.'" *Physics Letters A* 376 (45): 3466–3468. doi:10.1016/j.physleta.2012.10.010.

Oberkampf, William L. and Christopher J. Roy. 2010. *Verification and Validation in Scientific Computing.* 1st edition. New York: Cambridge University Press.

Oreskes, Naomi. 2004. "The Scientific Consensus on Climate Change." *Science* 306 (5702): 1686–1686.

——— 2007. "The Scientific Consensus on Climate Change: How Do We Know We're Not Wrong?" *Climate Change: What It Means for Us, Our Children, and Our Grandchildren*, 65–99. London: MIT Press. Reprinted in *Climate Modeling: Philosophical and Conceptual Issues,* edited by Elisabeth A. Lloyd and Eric Winsberg. 2018. Palgrave McMillan.

Oreskes, Naomi and Erik M. Conway. 2010. *Merchants of Doubt: How a Handful of Scientists Obscured the Truth on Issues from Tobacco Smoke to Global Warming.* Sydney, Australia: Bloomsbury Publishing USA.

Orzack, Steven Hecht and Elliott Sober. 1993. "A Critical Assessment of Levins's the Strategy of Model Building in Population Biology (1966)." *The Quarterly Review of Biology* 68 (4): 533–546.

Palmer, T. N. 2001. "A Nonlinear Dynamical Perspective on Model Error: A Proposal for Non-Local Stochastic-Dynamic Parametrization in Weather and Climate Prediction Models." *Quarterly Journal of the Royal Meteorological Society* 127 (572): 279–304. doi:10.1002/qj.49712757202.

Parker, Wendy S. 2010a. "Predicting Weather and Climate: Uncertainty, Ensembles and Probability." *Studies in History and Philosophy of Science Part B: Studies in History and Philosophy of Modern Physics* 41 (3): 263–272.

2010b. "Scientific Models and Adequacy-for-Purpose." *The Modern Schoolman* 87 (3/4): 285–293.

2011. "When Climate Models Agree: The Significance of Robust Model Predictions." *Philosophy of Science* 78 (4): 579–600. Reprinted in *Climate Modeling: Philosophical and Conceptual Issues,* edited by Elisabeth A. Lloyd and Eric Winsberg. 2018. Palgrave McMillan.

2014. "Values and Uncertainties in Climate Prediction, Revisited." *Studies in History and Philosophy of Science Part A* 46: 24–30.

Peterson, Thomas C. 2003. "Assessment of Urban Versus Rural In Situ Surface Temperatures in the Contiguous United States: No Difference Found." *Journal of Climate* 16 (18): 2941–2959. doi:10.1175/1520-0442(2003)016<2941:AOU VRI>2.0.CO;2.

Pirtle, Zachary, Ryan Meyer and Andrew Hamilton. 2010. "What Does It Mean When Climate Models Agree? A Case for Assessing Independence among General Circulation Models." *Environmental Science & Policy* 13 (5): 351–361.

Pugh, Charles and Maurício Matos Peixoto. 2008. "Structural Stability." *Scholarpedia* 3 (9): 4008. doi:10.4249/scholarpedia.4008.

RealClimateGroup. 2009. "The CRU Hack." RealClimate. www.realclimate.org/index.php/archives/2009/11/the-cru-hack/.

Reisch, George A. 2007. "From the 'Life of the Present' to the Icy Slopes of Logic": Logical Empiricism, the Unity of Science Movement, and the Cold War." In *The Cambridge Companion to Logical Empiricism,* edited by A. Richardson and T. Uebel, 58–87. Cambridge University Press.

Rosenberg, Stacy, Arnold Vedlitz, Deborah F. Cowman and Sammy Zahran. 2010. "Climate Change: A Profile of US Climate Scientists' Perspectives." *Climatic Change* 101 (3): 311–329.

Rougier, Jonathan and Michel Crucifix. 2018. "Uncertainty in Climate Science and Climate Policy." In *Climate Modeling: Philosophical and Conceptual Issues,* edited by Elisabeth A. Lloyd and Eric Winsberg. London: Palgrave McMillan.

Roy, Christopher J. 2005. "Review of Code and Solution Verification Procedures for Computational Simulation." *Journal of Computational Physics* 205 (1): 131–156.

Rudner, Richard. 1953. "The Scientist qua Scientist Makes Value Judgments." *Philosophy of Science* 20 (1): 1–6.

Santer, B. D., P. W. Thorne, L. Haimberger, K. E. Taylor, T. M. L. Wigley, J. R. Lanzante, S. Solomon et al. 2008. "Consistency of Modelled and Observed Temperature Trends in the Tropical Troposphere." *International Journal of Climatology* 28 (13): 1703–22. doi:10.1002/joc.1756.

Schmidt, Gavin A. 2008. "Of Buckets and Blogs." RealClimate. www.realclimate.org/index.php/archives/2008/06/of-buckets-and-blogs/.

2015. "NOAA Temperature Record Updates and the 'Hiatus'." RealClimate. www.realclimate.org/index.php/archives/2015/06/noaa-temperature-record-updates-and-the-hiatus/.

Schmidt, Gavin A. and Steven Sherwood. 2015. "A Practical Philosophy of Complex Climate Modelling." *European Journal for Philosophy of Science* 5 (2): 149–169.

Schupbach, Jonah N. 2016. "Robustness Analysis as Explanatory Reasoning." *The British Journal for the Philosophy of Science*, axw008, https://doi.org/10.1093/bjps/axw008.

Schupbach, Jonah N. and Jan Sprenger. 2011. "The Logic of Explanatory Power." *Philosophy of Science* 78 (1): 105–127.

Shackley, Simon, Peter Young, Stuart Parkinson and Brian Wynne. 1998. "Uncertainty, Complexity and Concepts of Good Science in Climate Change Modelling: Are GCMs the Best Tools?" *Climatic Change* 38 (2): 159–205. doi:10.1023/A:1005310109968.

Shewhart, Walter Andrew and William Edwards Deming. 1939. *Statistical Method from the Viewpoint of Quality Control*. North Chelmsford, MA: Courier Corporation.

Slingo, Julia and Tim Palmer. 2011. "Uncertainty in Weather and Climate Prediction." *Philosophical Transactions of the Royal Society A* 369 (1956): 4751–4767.

Smith, Leonard A. 2002. "What Might We Learn from Climate Forecasts?" *Proceedings of the National Academy of Sciences* 99 (suppl. 1): 2487–2492.

Smith, Peter. 1998. *Explaining Chaos*. Cambridge; New York: Cambridge University Press.

Solomon, Miriam. 2001. "*Social Empiricism*." MIT Press. https://mitpress.mit.edu/books/social-empiricism.

Steel, Daniel. 2010. "Epistemic Values and the Argument from Inductive Risk." *Philosophy of Science* 77 (1): 14–34.

2015. "Acceptance, Values, and Probability." *Studies in History and Philosophy of Science Part A* 53: 81–88.

Steele, Katie and Charlotte Werndl. 2013. "Climate Models, Calibration, and Confirmation." *The British Journal for the Philosophy of Science* 64 (3): 609–635.

Stegenga, Jacob. 2016. "Three Criteria for Consensus Conferences." *Foundations of Science* 21 (1): 35–49.

Stenhouse, Neil, Edward Maibach, Sara Cobb, Ray Ban, Andrea Bleistein, Paul Croft, Eugene Bierly, Keith Seitter, Gary Rasmussen and Anthony Leiserowitz. 2014. "Meteorologists' Views about Global Warming: A Survey of American Meteorological Society Professional Members." *Bulletin of the American Meteorological Society* 95 (7): 1029–1040.

Stern, N. H. 2007. *The Economics of Climate Change: The Stern Review*. Cambridge, UK: Cambridge University Press.

Stone, Dáithí A. and Reto Knutti. 2010. "Weather and Climate." In *Modelling the Impact of Climate Change on Water Resources*, edited by Fai Fung, Ana Lopez and Mark New, 4–33. John Wiley & Sons, Ltd. doi:10.1002/9781444324921. ch2.

Suárez, Mauricio. 2004. "An Inferential Conception of Scientific Representation." *Philosophy of Science* 71 (5): 767–779.

Suppes, Patrick. 1969. "Models of Data." In *Studies in the Methodology and Foundations of Science*, 24–35. Springer. http://link.springer.com/chapter/10.1007/978-94-017-3173-7_2.

Tebaldi, Claudia and Reto Knutti. 2007. "The Use of the Multi-Model Ensemble in Probabilistic Climate Projections." *Philosophical Transactions of the Royal Society of London A: Mathematical, Physical and Engineering Sciences* 365 (1857): 2053–2075.

"The Paris Agreement – Main Page." 2017. UNFCCC, accessed March 26. http://unfccc.int/paris_agreement/items/9485.php.

Thompson, Erica. 2013. "Modelling North Atlantic Storms in a Changing Climate." http://spiral.imperial.ac.uk/handle/10044/1/14730.

Tol, Richard. 2002a. Estimates of the Damage Costs of Climate Change. Part 1: Benchmark Estimates, *Environmental & Resource Economics* 21 (1): 47–73. doi:10.1023/A:1014500930521

    2002b. Estimates of the Damage Costs of Climate Change, Part II. Dynamic Estimates. *Environmental & Resource Economics* 21: 135–160. doi:10.1023/A:1014539414591.

Tucker, Aviezer. 2003. "The Epistemic Significance of Consensus." *Inquiry* 46 (4): 501–521. doi:10.1080/00201740310003388.

Verheggen, Bart, Bart Strengers, John Cook, Rob van Dorland, Kees Vringer, Jeroen Peters, Hans Visser and Leo Meyer. 2014. "Scientists' Views about Attribution of Global Warming." *Environmental Science & Technology* 48 (16): 8963–8971.

Weart, Spencer. 2010. "The Development of General Circulation Models of Climate." *Studies in History and Philosophy of Science Part B: Studies in History and Philosophy of Modern Physics* 41 (3): 208–217.

Weisberg, Michael. 2006. "Robustness Analysis." *Philosophy of Science* 73 (5): 730–742.

Wigley, T. M., V. Ramaswamy, J. R. Christy, J. R. Lanzante, C. A. Mears, B. D. Santer and C. K. Folland. 2006. "Temperature Trends in the Lower Atmosphere: Steps for Understanding and Reconciling Differences. Executive Summary." *US Climate Change Science Program, Synthesis and Assessment Product* 2 (11): 1–53.

Wilholt, Torsten. 2009. "Bias and Values in Scientific Research." *Studies in History and Philosophy of Science Part A* 40 (1): 92–101.

    2013. "Epistemic Trust in Science." *The British Journal for the Philosophy of Science* 64 (2): 233–253.

Wimsatt, W. C. 1994. "The Ontology of Complex Systems: Levels of Organization, Perspectives, and Causal Thickets." In Wimsatt, W.C., eds. *Re-engineering Philosophy for Limited Beings*, 193–240. Cambridge, MA: Harvard University Press.

2007. *Re-Engineering Philosophy for Limited Beings: Piecewise Approximations to Reality.* Cambridge, MA: Harvard University Press.

2011. *Robust Re-Engineering: A Philosophical Account?* Springer. http://link. springer.com/article/10.1007/s10539-011-9260-8/fulltext.html.

Winsberg, Eric. 2003. "Simulated Experiments: Methodology for a Virtual World." *Philosophy of Science* 70 (1): 105–125.

2006. "Handshaking Your Way to the Top: Simulation at the Nanoscale." *Philosophy of Science* 73 (5): 582–594.

2010. *Science in the Age of Computer Simulation.* University of Chicago Press.

2012. "Values and Uncertainties in the Predictions of Global Climate Models." *Kennedy Institute of Ethics Journal* 22 (2): 111–137.

2013. "Computer Simulations in Science." *Stanford Encyclopedia of Philosophy.* https://stanford.library.sydney.edu.au/entries/simulations-science/.

Winsberg, Eric and William Mark Goodwin. 2016. "The Adventures of Climate Science in the Sweet Land of Idle Arguments." *Studies in History and Philosophy of Science Part B: Studies in History and Philosophy of Modern Physics* 54: 9–17.

Winsberg, Eric, Bryce Huebner and Rebecca Kukla. 2014. "Accountability and Values in Radically Collaborative Research." *Studies in History and Philosophy of Science Part A* 46: 16–23.

Woodward, Jim. 2006. "Some Varieties of Robustness." *Journal of Economic Methodology* 13 (2): 219–240.

Worrall, John. 2014. "Prediction and Accommodation Revisited." *Studies in History and Philosophy of Science Part A* 45 (March): 54–61. doi:10.1016/ j.shpsa.2013.10.001.

# Index

259

Made in the USA
Las Vegas, NV
03 October 2021